ENERGY SCIENCE, ENGINEERING AND TECHNOLOGY

LARGE POWER TRANSFORMERS IN THE U.S. ELECTRIC GRID

ELEMENTS, ISSUES, AND SUBSTATION SECURITY

Energy Science, Engineering and Technology

Additional books in this series can be found on Nova's website under the Series tab.

Additional e-books in this series can be found on Nova's website under the e-book tab.

ENERGY SCIENCE, ENGINEERING AND TECHNOLOGY

LARGE POWER TRANSFORMERS IN THE U.S. ELECTRIC GRID

ELEMENTS, ISSUES, AND SUBSTATION SECURITY

VINCENT BECK
EDITOR

nova publishers
New York

NOTICE TO THE READER

Library of Congress Cataloging-in-Publication Data

ISBN: 978-1-63463-270-6

Published by Nova Science Publishers, Inc. † New York

CONTENTS

PREFACE

Chapter 1 – Large Power Transformers have long been a concern for the U.S. Electricity Sector, because the failure of a single unit can cause service interruption and lead to collateral damage, and there could be difficulties in quickly replacing them. This book assesses the procurement and supply environment of large power transformers (LPTs). Key industry sources have identified the limited availability of spare LPTs as a potential issue for critical infrastructure resilience in the United States, and both the public and private sectors have been undertaking a variety of efforts to address this concern. The following topics are examined in this book: characteristics and procurement of LPTs, including key raw materials and transportation; historical trends and future demands; global and domestic LPT suppliers; potential issues in the global sourcing of LPTs; and assessment of the risks facing LPTs.

Chapter 2 – In the United States, the electric power grid consists of over 200,000 miles of high-voltage transmission lines interspersed with hundreds of large electric power transformers. High voltage (HV) transformer units make up less than 3% of transformers in U.S. power substations, but they carry 60%-70% of the nation's electricity. Because they serve as vital nodes and carry bulk volumes of electricity, HV transformers are critical elements of the nation's electric power grid. HV transformers are also the most vulnerable to intentional damage from malicious acts. Recent security exercises, together with a 2013 physical attack on transformers in Metcalf, CA, have focused congressional interest on the physical security of HV transformers. They have also prompted new grid security initiatives by utilities and federal regulators. Legislative proposals, notably the Grid Reliability and Infrastructure Defense Act (H.R. 4298 and S. 2158), would expand these efforts by strengthening federal authority to secure the U.S. grid.

For more than 10 years, the electric utility industry and government agencies have engaged in a number of initiatives to secure HV transformers from physical attack and to improve recovery in the event of a successful attack. These initiatives include coordination and information sharing, spare equipment programs, security standards, grid security exercises, and other measures. There has been some level of physical security investment and an increasing refinement of voluntary grid security practices across the electric power sector for at least the last 15 years. Several major transmission owners have recently announced significant new initiatives specifically to improve the physical security of critical transformer substations in light of the Metcalf attack.

On March 7, 2014, the Federal Energy Regulatory Commission (FERC) ordered the North American Electric Reliability Corporation (NERC) to submit to the Commission new reliability standards requiring certain transmission owners "to take steps or demonstrate that they have taken steps to address physical security risks and vulnerabilities related to the reliable operation" of the power grid. In its order, FERC states that physical security standards are necessary because "the current Reliability Standards do not specifically require entities to take steps to reasonably protect against physical security attacks." According to FERC's order, the new reliability standards will require grid owners to perform risk assessments to identify their critical facilities, evaluate potential threats and vulnerabilities, and implement security plans to protect against attacks.

There is widespread agreement among state and federal government officials, utilities, and manufacturers that HV transformers in the United States are vulnerable to terrorist attack, and that such an attack potentially could have catastrophic consequences. But the most serious, multi-transformer attacks would require acquiring operational information and a certain level of sophistication on the part of potential attackers. Consequently, despite the technical arguments, without more specific information about potential targets and attacker capabilities, the true vulnerability of the grid to a multi-HV transformer attack remains an open question. Incomplete or ambiguous threat information may lead to inconsistency in physical security among HV transformer owners, inefficient spending of limited security resources at facilities that may not really be under threat, or deployment of security measures against the wrong threat.

As the electric power industry and federal agencies continue their efforts to improve the physical security of critical HV transformer substations, Congress may consider several key issues as part of its oversight of the sector:

identifying critical transformers, confidentiality of critical transformer information, adequacy of HV transformer protection, quality of federal threat information, and recovery from HV transformer attacks.

In: Large Power Transformers
Editor: Vincent Beck

ISBN: 978-1-63463-270-6
© 2015 Nova Science Publishers, Inc.

Chapter 1

LARGE POWER TRANSFORMERS AND THE U.S. ELECTRIC GRID[*]

Office of Electricity Delivery and Energy Reliability

EXECUTIVE SUMMARY

This study updates the initial June 2012 study under the same title, *Large Power Transformers and the U.S. Electric Grid.* In this report, the Office of Electricity Delivery and Energy Reliability, U.S. Department of Energy (DOE) assessed the procurement and supply environment of large power transformers (LPTs).[1] LPTs have long been a concern for the U.S. Electricity Sector, because the failure of a single unit can cause temporary service interruption and lead to collateral damage, and it could be difficult to quickly replace it. Key industry sources have identified the limited availability of spare LPTs as a potential issue for critical infrastructure resilience in the United States, and both the public and private sectors have been undertaking a variety of efforts to address this concern. Therefore, DOE examined the following topics in this report: characteristics and procurement of LPTs, including key raw materials and transportation; historical trends and future demands; global and domestic LPT suppliers; potential issues in the global sourcing of LPTs; and assessment of the risks facing LPTs.

---

[*] This is an edited, reformatted and augmented version of an updated report issued by the U.S. Department of Energy, April 2014.

LPTs are custom-designed equipment that entails a significant capital expenditure and a long lead time due to an intricate procurement and manufacturing process. Although prices vary by manufacturer and by size, an LPT can cost millions of dollars and weigh between approximately 100 and 400 tons (or between 200,000 and 800,000 pounds). The procurement and manufacturing of LPTs is a complex process that includes prequalification of manufacturers, a competitive bidding process, the purchase of raw materials, and special modes of transportation due to its size and weight. The result is the possibility of an extended lead time that could stretch beyond 20 months if the manufacturer has difficulty obtaining certain key parts or materials. Two raw materials—copper and electrical steel—account for more than half of the total cost of an LPT. Special grade electrical steel is used for the core of a power transformer and is critical to the efficiency and performance of the equipment; copper is used for the windings. In recent years, the price volatility of these two commodities in the global market has affected the manufacturing condition and procurement strategy for LPTs.

The rising global demand for copper and electrical steel can be partially attributed to the increased power and transmission infrastructure investment in growing economies, as well as the replacement market for aging infrastructure in developed countries. The United States is one of the world's largest markets for power transformers and holds the largest installed base of LPTs, and this installed base is aging. The average age of installed LPTs in the United States is approximately 38 to 40 years, with 70 percent of LPTs being 25 years or older. While the life expectancy of a power transformer varies depending on how it is used, aging power transformers are potentially subject to an increased risk of failure.

Since the late 1990's, the United States has experienced an increased demand for LPTs; however, despite the growing need, the United States has a limited domestic capacity to produce LPTs. In 2010, six power transformer manufacturing facilities existed in the United States, and together, they met approximately 15 percent of the Nation's demand for power transformers of a capacity rating greater than or equal to 60 megavolt-amperes (MVA). Although the exact statistics are unavailable, global power transformer supply conditions indicate that the Nation's reliance on foreign manufacturers was even greater for extra high-voltage (EHV) power transformers with a maximum voltage rating greater than or equal to 345 kilovolts (kV).

However, the domestic production capacity of LPTs in the United States has seen some improvements. Since April 2010, four new or expanded facilities have begun producing LPTs in the United States, including: Efacec's first U.S. transformer plant, which began production in Rincon, Georgia, in April 2010; Hyundai Heavy Industries' new manufacturing facility, which was inaugurated in Montgomery, Alabama, in November 2011; SPX Transformer Solution's facility in Waukesha, Wisconsin, which completed expansion in April 2012; and Mitsubishi's new power transformer plant in Memphis, Tennessee, which became operational in April 2013.

The upward trend of transmission infrastructure investment in the United States since the late 1990s is one of the key drivers for the recent addition of domestic manufacturing capacity for power transformers. Power transformers are globally-traded equipment, and the demand for this machinery is forecasted to continue to grow at a compound annual growth rate of three percent to seven percent in the United States according to industry sources. In addition to the need for the replacement of aging infrastructure, the United States has a demand for transmission expansion and upgrades to accommodate new generation connections and maintain electric reliability.

While global procurement has been a common practice for many utilities to meet their growing need for LPTs, there are several challenges associated with it. Such challenges include: the potential for an extended lead time due to unexpected global events or difficulty in transportation; the fluctuation of currency exchange rates and material prices; and cultural differences and communication barriers. The utility industry is also facing the challenge of maintaining an experienced in-house workforce that is able to address procurement and maintenance issues.

The U.S. electric power grid is one of the Nation's critical life-line functions on which many other critical infrastructure depend, and the destruction of this infrastructure can have a significant impact on national security and the U.S. economy. The electric power infrastructure faces a wide variety of possible threats, including natural, physical, cyber, and space weather. While the potential effect of these threats on the infrastructure is uncertain, public and private stakeholders in the energy industry are considering and developing a variety of risk management strategies to mitigate the effects. This DOE report updates the prior 2012 study and includes the following additional discussions:

- Updated information about global electrical steel supply conditions;
- The increased domestic production of LPTs resulting from four new or expanded plants;
- The historical assessment of risks to power transformers by an insurance firm; and

- New government and industry efforts to augment risk management options for critical electricity infrastructure, including power transformers.

Through these and the assessment of the manufacturing and supply issues related to LPTs, this report provides information to help the industry's continuous efforts to build critical energy infrastructure resilience in today's complex, interdependent global economy.

1. INTRODUCTION

1.1. The Focus of the Study

In today's dynamic, intersected global economies, understanding market characteristics and securing the supply basis of critical equipment, such as large power transformers (LPTs),[2] becomes increasing imperative for maintaining the resilience of the Electricity Sector.[3] The purpose of this report is to provide information to decision-makers in both public and private sectors about the country's reliance on foreign-manufactured LPTs and potential supply issues.

The Office of Electricity Delivery and Energy Reliability, U.S. Department of Energy (DOE), as part of its ongoing efforts to enhance the resilience of the Nation's critical infrastructure, assessed the manufacturing and supply conditions of LPTs in this report.[4] Power transformers have long been a concern for the U.S. Electricity Sector.[5] The failure of a single unit could result in temporary service interruption and considerable revenue loss, as well as incur replacement and other collateral costs. Should several of these units fail at the same time, it will be challenging to quickly replace them.

LPTs are special-ordered machineries that require highly skilled workforces and state-of-the-art manufacturing equipment and facilities. The installation of LPTs entails not only significant capital expenditures but also a long lead time due to the intricate manufacturing processes, including the securing of raw materials. As a result, asset owners and operators invest considerable resources to monitor and maintain LPTs, as failure to replace aging LPTs could present potential concerns, including increased maintenance costs, equipment failures, and unexpected power failures.

Therefore, this report examines the following:

- Classification and physical characteristics of LPTs;
- Power transformer procurement and manufacturing processes;
- Supply sources and price variability of two raw commodities—copper and electrical steel;
- Global and domestic power transformer market and manufacturing conditions;
- Key global suppliers of LPTs to the United States;
- Potential challenges in the global sourcing of power transformers; and
- Assessment of potential risks and threats to power transformers.

1.2. Background

As applied to infrastructure, the concept of resilience embodies "the ability to adapt to changing conditions and prepare for, withstand, and rapidly recover from disruption."[6] The resilience strategy is a cornerstone of the U.S. national policy as adopted in the 2013 *NIPP: Partnering for Critical Infrastructure Security and Resilience*, the 2012 *National Strategy for Global Supply Chain Security,* and the 2010 *Energy Sector-Specific Plan.*[7] Broadly defined as the ability to withstand and recover from adversity, resilience is also increasingly applied to broad technical systems and social context.

The Electricity Sector has long embraced resilience as part of continuity of operations planning, risk management, and systems reliability. These practices have been so well ingrained in the operation of the electric grid that utility owners and operators often do not think of their practices as "resilience."[8] Although reliability and redundancy are built into the system, the electricity industry identified that the limited domestic manufacturing capacity of high-voltage power transformers could present a potential supply issue in the event that many LPTs failed simultaneously.[9]

This concern was brought to light in the Electricity Sector's High-Impact Low-Frequency (HILF) Risk Workshop in November 2009 and in a subsequent report entitled, *High Impact, Low Frequency Event Risk to the North American Bulk Power System* in the following year.[10] Co-sponsored by DOE and the North American Electrical Reliability Corporation (NERC), the HILF Risk Workshop examined selected severe impact risks to the Nation's electrical grid. The workshop considered four risk scenarios concerning the Electricity Sector, including a severe geomagnetic disturbance (GMD) or electromagnetic pulse (EMP) event that damaged a difficult-to-replace generating station and substation equipment causing a cascading effect on the system.

In November 2010, NERC released the Critical Infrastructure Strategic Roadmap to provide a framework on how to address some of the severe impact risks identified in the HILF report.[11] The Roadmap provided recommendations on how to enhance electricity reliability and resilience from an all-hazards perspective and suggested direction for the Electricity Sector. Specifically, the Roadmap advised Electricity Sector entities to consider a full spectrum of risk management elements to address severe impact risks—planning, prevention, mitigation, and recovery.[12] In accordance with the Roadmap, both the public and private sectors of the Electricity Sector have undertaken a variety of activities that consider these risk management elements, including the recent development of the NERC Reliability Standards to protect the bulk power system from the effects of GMD.

This DOE report supplements the sector's ongoing resilience efforts, specifically the prevention and recovery elements, through an examination of the supply chain of LPTs.

1.3. Scope and Definition of Large Power Transformers

Throughout this report, the term LPT is broadly used to describe a power transformer with a maximum nameplate rating of 100 megavolt-amperes (MVA) or higher unless otherwise noted. However, it should be noted that there is no single, absolute industry definition or criterion for what constitutes an LPT and that additional specifications are often used to describe different classes of LPTs.

The size of a power transformer is determined by the primary (input) voltage, the secondary (output) voltage, and the load capacity measured by MVA. Of the three, the capacity rating, or the amount of power that can be transferred, is often the key parameter rather than the voltage.[13]

In addition to the capacity rating, voltage ratings are often used to describe different classes of power transformers, such as extra high voltage (EHV), 345 to 765 kilovolts (kV); high voltage, 115 to 230 kV; medium voltage, 34.5 to 115 kV; and distribution voltage, 2.5 to 35 kV.[14] A power transformer with a capacity rating greater than or equal to 100 MVA typically has a voltage rating of greater than or equal to 115 kV on the high side; therefore, an LPT with a capacity rating of 100 MVA or greater can have a transmission voltage class of medium, high, or extra high (greater than or equal to115 kV).

There are considerable differences in the definitions used to describe an LPT, including the transmission voltage classifications shown in Table 1. This report derived the criterion of an LPT from the following sources:

- In a 2006 DOE study entitled, *Benefits of Using Mobile Transformers and Mobile Substations for Rapidly Restoring Electrical Service*, LPTs were described as "high-power transformers . . . with a rating over 100 MVA."[15]

- A 2011 report from an antidumping investigation by the United States International Trade Commission (USITC) established LPTs as "large liquid dielectric power transformers having a top power handling capacity greater than or equal to 60,000 kilovolt amperes (60 megavolt amperes), whether assembled or unassembled, complete or incomplete."[16]

- The 2011 NERC Spare Equipment Database Task Force Report defined LPTs as follows:[17]

  - Transmission Transformers: The low voltage side is rated 100 kV or higher and the maximum nameplate rating is 100 MVA or higher.

  - Generation Step-up Transformers: The high voltage side is 100 kV or higher and the maximum nameplate rating is 75 MVA or higher.

2. POWER TRANSFORMER CLASSIFICATION

2.1. Power Transformers in the Electric Grid

North America's electricity infrastructure represents more than $1 trillion U.S. dollars in asset value and is one of the most advanced and reliable systems in the world. The U.S. bulk grid consists of approximately 390,000 miles of transmission lines, including more than 200,000 miles of high-voltage lines, connecting to more than 6,000 power plants.[18] Power transformers are a critical component of the transmission system, because they adjust the electric voltage to a suitable level on each segment of the power transmission from generation to the end user. In other words, a power transformer steps up the voltage at generation for efficient, long-haul transmission of electricity and steps it down for distribution to the level used by customers.[19] Power transformers are also needed at every point where there is a change in voltage in power transmission to step the voltage either up or down. Figure 1 illustrates a simplified arrangement of the U.S. electric grid system.

2.2. Physical Characteristics of Large Power Transformers

An LPT is a large, custom-built piece of equipment that is a critical component of the bulk transmission grid. Because LPTs are very expensive and tailored to customers' specifications, they are usually neither interchangeable with each other nor produced for extensive spare inventories.[20] According to an industry source, approximately 1.3 transformers are produced for each transformer design. Figure 2 illustrates a standard core-type LPT and its major internal components.

Although LPTs come in a wide variety of sizes and configurations, they consist of two main active parts: the core, which is made of high-permeability, grain-oriented, silicon electrical steel, layered in pieces; and windings, which are made of copper conductors wound around the core, providing electrical input and output. Two basic configurations of core and windings exist—the core form and the shell form. In the usual shell-type power transformer, both primary and secondary windings are on one leg and are surrounded by the core, whereas in a core-type power transformer, cylindrical windings cover the core legs. Shell-form LPTs typically use more electrical steel for the core and are more resilient to short-circuits in the transmission systems and are frequently used in industrial applications.[21] The core and windings are contained in a rectangular, mechanical frame called the "tank." Other parts include bushings, which connect LPTs to transmission lines, as well as tap changers, power cable connectors, gas-operated relays, thermometers, relief devices, dehydrating breathers, oil level indicators, and other controls.[22]

Table 1. Transmission Voltage Classes

Class	Voltage Ratings (kV)
Medium Voltage	34.5, 46, 69, 115/138
High Voltage	115/138, 161, 230
Extra High Voltage	345, 500, 765

Source: DOE, 2006; see footnote 13. Modified based on industry review.

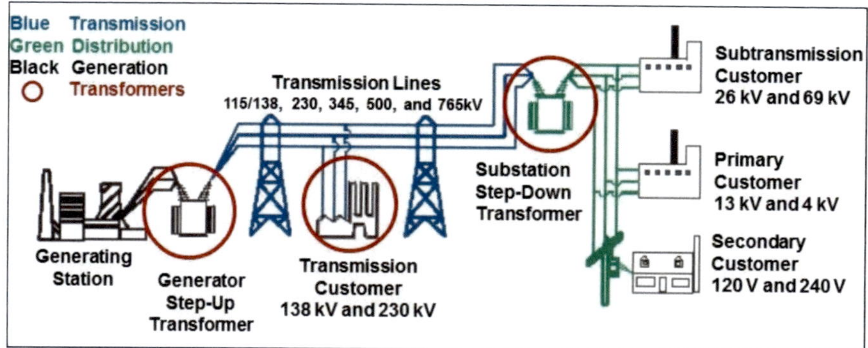

Source: DOE, 2006; see footnote 13. Modified based on industry review.

Figure 1. Electric Power Grid Representation.

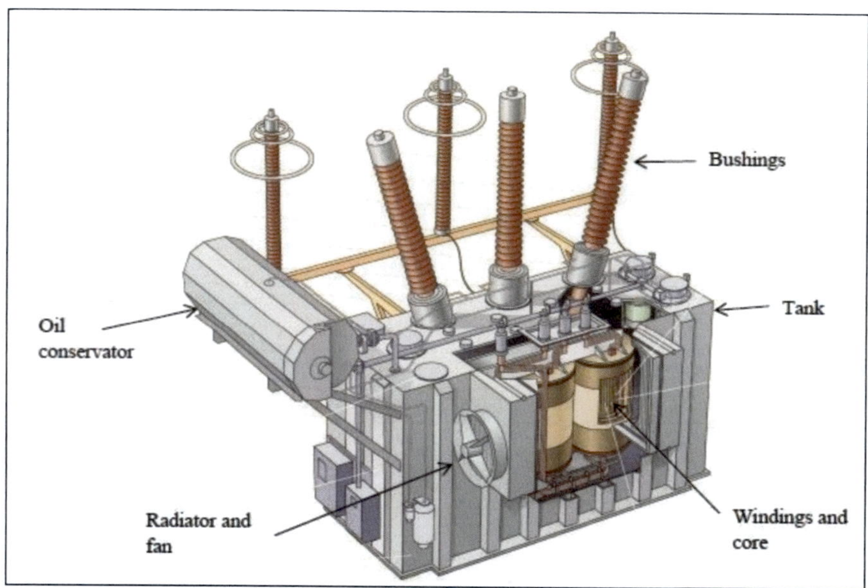

Source: "Liquid-Filled Power Transformers," ABB http://www05.abb.com/global/
scot/scot252.nsf/veritydisplay/299a52373c3fd0e6c12578be003a476f/$file/pptr_m
pt_brochure_2406pl170-w1- en.pdf (accessed March 31, 2014).

Figure 2. Core-Type Large Power Transformer Showing Major Internal Components.

Power transformer costs and pricing vary by manufacturer, market condition, and location of the manufacturing facility. In 2010, the approximate cost of an LPT with an MVA rating between 75 MVA and 500 MVA was estimated to range from $2 million to $7.5 million in the United States;

however, these estimates were Free on Board (FOB) factory costs, exclusive of transportation, installation, and other associated expenses, which generally add 25 percent to 30 percent to the total cost (see Table 2).[23] Raw materials, particularly copper and electrical steel, are a significant factor in power transformer prices. Transportation is also an important element of the total LPT cost because an LPT can weigh as much as 410 tons (820,000 pounds (lb)) and often requires long-distance transport.

Table 2. Estimated Magnitude of Large Power Transformers in 2011

Voltage Rating (Primary-Secondary)	Capability MVA Rating	Approximate Price	Approximate Weight and Dimensions
Transmission Transformer			
Three Phase			
230–115kV	300	$2,000,000	170 tons (340,000 lb) 21ft W–27ft L–25ft H
345–138kV	500	$4,000,000	335 tons (670,000 lb) 45ft W–25ft L–30ft H
765–138kV	750	$7,500,000	410 tons (820,000 lb) 56ft W–40ft L–45ft H
Single Phase			
765–345kV	500	$4,500,000	235 tons (470,000 lb) 40ft W–30ft L–40ft H
Generator Step-Up Transformer			
Three Phase			
115–13.8kV	75	$1,000,000	110 tons (220,000 lb) 16ft W–25ft L–20ft H
345––13.8kV	300	$2,500,000	185 tons (370,000 lb) 21ft W–40ft L–27ft H
Single Phase			
345–22kV	300	$3,000,000	225 tons (450,000 lb) 35ft W–20ft L–30ft H
765–26kV	500	$5,000,000	325 tons (650,000 lb) 33ft W–25ft L–40ft H

Note: Prices are FOB factory and do not include taxes, transportation, special features and accessories, special testing (short-circuit, etc.), insulating oil, field installation, and/or optional services. The total installed cost is estimated to be about 25 percent to 30 percent higher.

Source: "Special Report: Spare Equipment Database System," NERC, 2011; see footnote 17.

LPTs require substantial capital and a long-lead time (in excess of six months) to manufacture, and its production requires large crane capacities, ample floor space, and adequate testing and drying equipment. The following section provides further discussions on the production processes and requirements of LPTs, including transportation and key raw commodities.

3. LARGE POWER TRANSFORMER PROCUREMENT AND MANUFACTURING PROCESS

3.1. Overview

This section provides an overview of key steps in the procurement and manufacturing process of an LPT, including bidding, production, and transportation. This overview is then followed by a discussion of key raw materials—electrical steel and copper—which are integral to LPTs. The several distinct steps and procedures, as well as the estimated lead time for each step required in power transformer manufacturing and procurement, are illustrated in Figure 3.

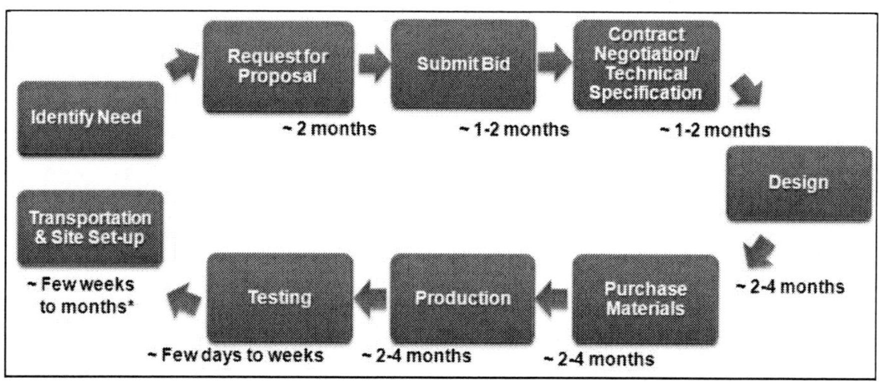

Note: This figure illustrates an optimal flow of the manufacturing process and the estimated lead time, which can extend beyond the estimated time frame shown.
* Variable depending on distance and logistical issues.
Source: USITC and industry estimate.

Figure 3. Large Power Transformer Procurement Process and Estimated Lead Time.

3.1.1. Prequalification of Manufacturers

As discussed in Section 2, LPTs are custom-made equipment that incurs significant capital costs. Utilities generally procure LPTs through a competitive bidding process, in which all interested producers must prequalify to be eligible to bid. Prequalification is a lengthy process that can take several years.[24] A typical qualification process includes an audit of production and quality processes, verification of certain International Organization for Standardization (ISO) certifications, and inspection of the manufacturing environment. This process can often be rigorous and costly to purchasers; however, it is an important step, because the manufacturing environment and capability can significantly affect the reliability of the product, especially of high-voltage power transformers.[25]

3.1.2. Bidding Process

A standard bidding process is initiated by a purchaser, who sends commercial specifications to qualified LPT producers. The producers then design LPTs to meet the specifications, estimate the cost, and submit a bid to the purchaser. The bids not only include the power transformer, but also services such as transportation, installation, and warranties.[26] Except for a few municipalities, most U.S. utilities do not announce the amount of the winning bid or the identity of the winning bidder. The winning bidder is notified, and bid terms normally require that the results be kept confidential by all parties involved.[27]

3.1.3. Production

The typical manufacturing process of an LPT consists of the following steps:[28]

1. **Engineering and design:** LPT design is complex, balancing the costs of raw materials (copper, steel, and cooling oil), electrical losses, manufacturing labor hours, plant capability constraints, and shipping constraints.
2. **Core building:** The core is the most critical component of an LPT, which requires a highly-trained and skilled workforce and cold-rolled, grain-oriented (CRGO) laminated electrical steel.
3. **Windings production and assembly of the core and windings:** Windings are predominantly copper and have an insulating material.

4. **Drying operations:** Excess moisture must be removed from the core and windings because moisture can degrade the dielectric strength of the insulation.

5. **Tank production:** A tank must be completed before the winding and core assembly finish the drying phase so that the core and windings do not start to reabsorb moisture.

6. **Final assembly of the LPT:** The final assembly must be done in a clean environment; even a tiny amount of dust or moisture can deteriorate the performance of an LPT.

7. **Testing:** Testing is performed to ensure the accuracy of voltage ratios, verify power ratings, and determine electrical impedances.

In the manufacturing process, certain parts can be produced either at the transformer plant or at another vendor or subsidiary location, depending on how vertically integrated the particular plant is and whether the plant has the necessary tools and capabilities, as well as for economic reasons.[29]

3.1.4. Lead Time

In 2010, the average lead time between a customer's LPT order and the date of delivery ranged from five to 12 months for domestic producers and six to 16 months for producers outside the United States.[30] The LPT market is characterized as a cyclical market with a correlation between volume, lead time, and price. In other words, the average lead time can increase when the demand is high, up to 18 to 24 months.[31] This lead time could extend beyond 20 months and up to five years in extreme cases if the manufacturer has difficulty obtaining any key inputs, such as bushings and other key raw materials, or if considerable new engineering is needed.[32] An industry source noted that high-voltage (HV) bushings often have a long lead time, extending up to five months. Another industry source added that HV bushings are usually customized for each power transformer and there are limited bushing manufacturers in the United States. Manufacturers must also secure supplies of specific raw materials or otherwise they could endure an extended lead time.[33]

Once completed, a power transformer is disassembled for transport, including the removal of oil, radiators, bushings, convertors, arrestors, and so forth. The proper transportation of a power transformer and its key parts is critical to ensuring the high reliability of the product and minimizing the period for onsite installation.

3.1.5. Transportation

Transporting an LPT is challenging—its large dimensions and heavy weight pose unique requirements to ensure safe and efficient transportation. Current road, rail, and port conditions are such that transportation is taking more time and becoming more expensive.[34] Although rail transport is most common, LPTs cannot be transferred over normal railcars, because they cannot be rolled down a hill or bumped into other rail cars, which can damage the power transformer. This is because the heaviest load a railroad normally carries is about 100 tons, or 200,000 lb, whereas an LPT can weight two to three times that amount.[35]

A specialized railroad freight car known as the Schnabel railcar is used to transport extremely heavy loads and accommodate height via railways. Figure 4 shows LPTs being transported on the road and on a Schnabel car. There are a limited number of Schnabel cars available worldwide, with only about 30 of them in North America.[36] Certain manufacturers operate a Schnabel car rental program and charge up to $2,500 per day for the rental in addition to other applicable fees.[37] Access to a railroad is also becoming an issue in certain areas due to the closure, damage, or removal of rail lines. A German machine called the Goldhofer, which "looks like a caterpillar with 144 tires and features a hydraulic system" to handle the heavy weight, is another mode of transportation used on the road.[38]

When an LPT is transported on the road, it requires obtaining special permits and routes from the department of transportation of each state on the route of the LPT being transported. According to an industry source, obtaining these special permits can require an inspection of various infrastructure (e.g., bridges), which can add delay. In addition, transporting LPTs on the road can require temporary road closures due to traffic issues, as well as a number of crew and police officers to coordinate logistics and redirect traffic. The transport modular shown in Figure 4 is 70 feet long with 12 axles and 192 wheels, and occupies two lanes of traffic.

Logistics and transportation accounted for approximately three percent to 20 percent of the total cost of an LPT for both domestic and international producers.[39] While important, this is less significant than the cost of raw materials and the potential sourcing concerns surrounding them. The next section describes some of the issues concerning raw materials vital to LPT manufacturing.

Note: Workers move wires, lights, and poles to transport a 340-ton power transformer, causing hours of traffic delay.
Source: Pittsburgh Live News, December 2011.

Note: Schnabel Car transporting an LPT.
Source: ABB.

Figure 4. Transport of Large Power Transformers.

3.2. Raw Materials Used in Large Power Transformers

The main raw materials needed to build power transformers are copper conductors, silicon iron/steel, oil, and insulation materials. The cost of these raw materials is significant, accounting for well over 50 percent of the total cost of a typical LPT. Specifically, manufacturers have estimated that the cost of raw materials accounted for 57 percent to 67 percent of the total cost of LPTs sold in the United States between 2008 and 2010.[40] Of the total material cost, about 18 percent to 27 percent was for copper and 22 percent to 24 percent was for electrical steel.[41] For this reason, this section examines the issues surrounding the supply chain and price variability of the two key raw materials used in LPTs—copper and electrical steel.

3.2.1. Electrical Steel and Large Power Transformers

The electrical steel used in power transformer manufacture is a specialty steel tailored to produce certain magnetic properties and high permeability. A special type of steel called cold-rolled grain-oriented electrical steel (hereinafter refer to as "electrical steel") makes up the core of a power transformer. Electrical steel is the most critical component that has the greatest impact on the performance of the power transformer, because it is designed to provide low core loss and high permeability, which are essential to efficient and economical power transformers.

Electrical steel is produced in different levels of magnetic permeability: conventional and high- permeability. Conventional products are available in various grades from M-2 through M-6, with thickness and energy loss increasing with each higher number (see Figure 5).[42] High-permeability product allows a transformer to operate at a higher level of flux density[43] than conventional products, thus permitting a transformer to be smaller and have lower operating losses.[44] The quality of electrical steel is measured in terms of loss of electrical current flowing in the core. In general, core losses are measured in watts per kilogram (W/kg), and the thinner the material, the better the quality.[45] An industry source noted that an electrical steel grade of M3 or better is typically used in LPTs to minimize core loss.

The quality or degree of loss that is acceptable can vary depending on the primary use of the power transformer—whether it is located on a site that carries a moderate or heavy load. In other words, the grade of electrical steel used for the power transformer core varies depending on how highly the utility values losses.[46] Recently, due to environmental protection requirements, energy savings and minimizing core loss in power transformers are becoming important.

Figure 5 present the average annual prices of grain-oriented electrical steel in the United States between 2006 and 2011. The average annual prices of electrical steel ranged from $1.20 to $2.20 per pound between 2006 and 2011, with peak prices occurring in 2008. According to an industry source, the price of electrical steel has been recorded as high as $2.80 per pound (lb). As a reference, approximately 170,000 to 220,000 lb of core steel is needed in a power transformer with a capacity rating between 300 and 500 MVA.[47]

As shown in figure 5, the prices of electrical steel have been volatile over the time period from 2006 to 2008 and have generally declined from their 2008 highs. These trends can be attributed to the following factors:[48]

- The continued increase in global demand for grain oriented electrical steel, particularly in China and India;
- Higher raw material prices to the core steel manufacturers (e.g., iron ore, coking coal, scrap steel) and higher processing energy costs;
- The low value of the U.S. dollar, (the low value increases the cost of imported steel and encourages domestic suppliers to export); and
- The onset of the global economic recession and financial meltdown in late 2008.

3.2.2. Global Electrical Steel Suppliers

The availability of electrical steel supply sources worldwide is limited. In 2013, there were only two domestic producers—AK Steel and Allegheny Ludlum. In addition to the two domestic producers, there were eleven major international companies producing grain oriented electrical steel. However, only a limited number of producers worldwide are capable of producing the high-permeability steel that is generally used in LPT cores. AK Steel is the only domestic producer of the high-permeability, domain-refined (laser-scribed) core steel used in high- efficiency stacked cores.

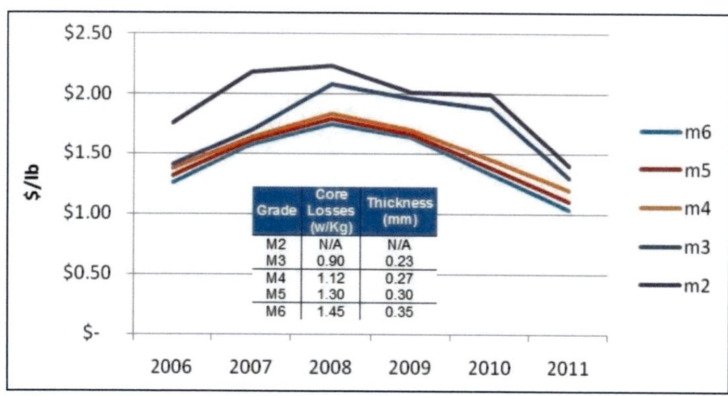

Grade	Core Losses (w/Kg)	Thickness (mm)
M2	N/A	N/A
M3	0.90	0.23
M4	1.12	0.27
M5	1.30	0.30
M6	1.45	0.35

Source: DOE EERE, 2013; see footnote 48.

Figure 5. Average Annual Prices of Grain-Oriented Electrical Steel in the United States from 2006 to 2011.

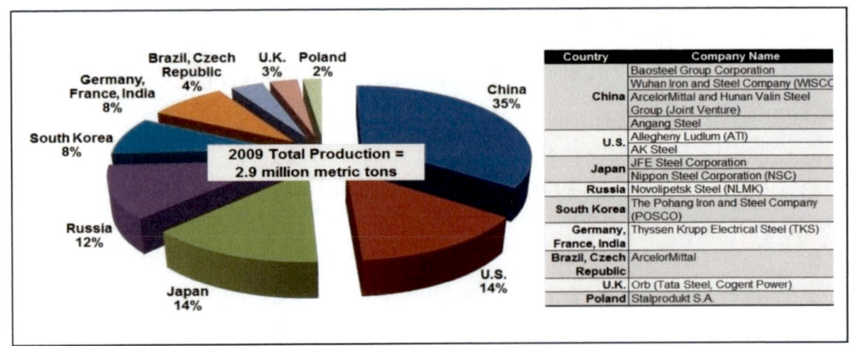

Source: DOE EERE, 2013; see footnote 48.

Figure 6. World Installed Production of Electrical Steel in 2009.

Figure 6 represents electrical steel production by country in 2009, as well as the names of producers from each country. In 2009, China's four companies produced 35 percent of the world's electrical steel, the majority of which were consumed domestically. Conversely, Japan produced 14 percent of the world's electrical steel mainly for the purpose of export. The two U.S. producers accounted for 14 percent of the world's electrical steel production.

According to the USITC, in 2012, a total of 1.5 million metric tons of electrical steel were exported around the world, and exports from three countries—Japan, Russia, and South Korea—accounted for more than half of that total (see Figure 7).[49] While China was the largest producer of electrical steel, it contributed only two percent of the total global export in 2012. Japan was the largest exporter of electrical steel with 27 percent, followed by Russia and Korea, each exporting 15 percent and 10 percent of total global electrical steel in 2012, respectively. See "Appendix B. Global Electrical Steel Manufacturer Profiles," for additional information on the electrical steel producers.

3.2.3. Variability of Commodity Prices

Since 2004, the global commodity market has experienced price fluctuations for both steel and copper, driven up largely by the demand from emerging economies. The global consumption of these metals is expected to continue to rise in the next decade as supply conditions tighten, leading to increased worries about the future price movement of these key commodities. The average price of copper more than quadrupled between 2003 and 2013, costing more than $4.27 per pound by 2011 (see Figure 8). While the price of steel shown in this figure does not present the electrical steel used to manufacture LPTs, it does shed some light on the volatility of steel commodity prices worldwide.

In 2012, China was the single largest buyer of steel in the world, consuming more than 45 percent of the world's total steel consumption of 1,413 million metric tons that year.[50] Although China's primary need for steel is in the construction sector, China also has a significant demand for power transmission infrastructure. China's and India's demands for steel, including high- efficiency, grain-oriented steel, are expected to continue to affect the availability and price of steel and copper to the rest of the world.[51]

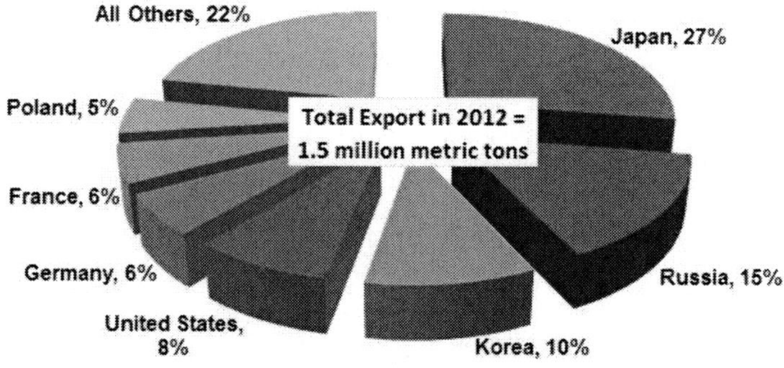

Source: USITC, 2013.

Figure 7. Global Exports of Electrical Steel by Reporting Countries in 2012.

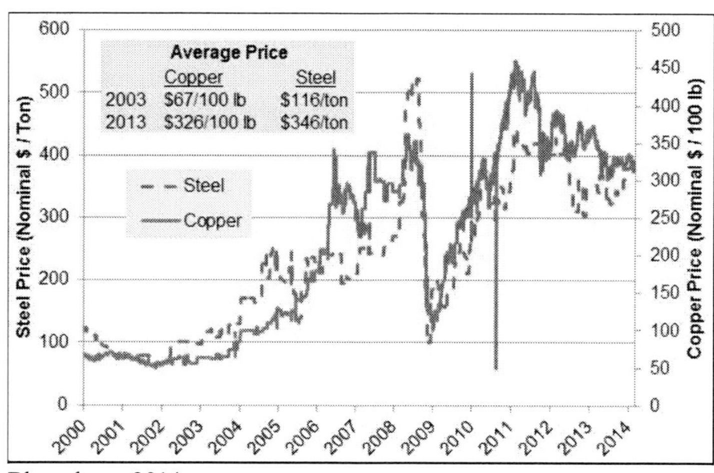

Source: Bloomberg, 2014.

Figure 8. Historical Copper and Steel Price Variability from 2000 to 2013.

Faced with the price volatility of raw materials and increased global demand for LPTs in recent years, the industry's procurement teams have revamped their sourcing strategies. According to a sourcing analyst from a major U.S. electric utility, a procurement strategy companies are increasingly using is multiyear "blanket agreements" in which they lock in volumes and price points for power transformers.[52] Blanket agreements are long-term alliances between an investor- owned utility and a specific LPT supplier; the utility selects and locks in with one manufacturer to provide it with LPTs for a period of two to five years. These agreements benefit the utility because once it provides specifications and buys one power transformer from the supplier, additional LPTs can be produced and shipped more rapidly. Although industry executives suggest that sales based on these alliances account for a significant share of LPT sales in the United States, the actual percentage of sales based on these agreements is unavailable.[53]

4. POWER INFRASTRUCTURE INVESTMENT TRENDS

4.1. Global Power Generation Capacity

In 2013, the world had more than five trillion watts of power generation capacity, which was growing at an annual rate of two percent. China and the United States had the largest generation capacity, with each holding about 21 percent and 20 percent of the world's total installed capacity, respectively.[54]

According to the U.S. Energy Information Administration's (EIA) 2013 *International Energy Outlook*, the world will add 2,840 gigawatts (GW) of new capacity by 2040, a majority of which will be installed in non-OECD (Organization for Economic Cooperation and Development) countries.[55] China's generating capacity is forecasted to increase by approximately three percent annually through 2040, while the U.S. capacity is expected to rise at approximately one percent annually during the same period (see Figure 9).[56] By 2040, China's installed power generation capacity will surpass that of the United States and will hold a quarter of the world's total capacity of 8,254 GW, becoming the world's largest power generating capacity holder. As a general rule, the addition of electricity generation capacity spurs investment in not only power generation infrastructure but also in transmission infrastructure, including transmission line equipment and power transformers, which are necessary to transmit power from generators to end users.

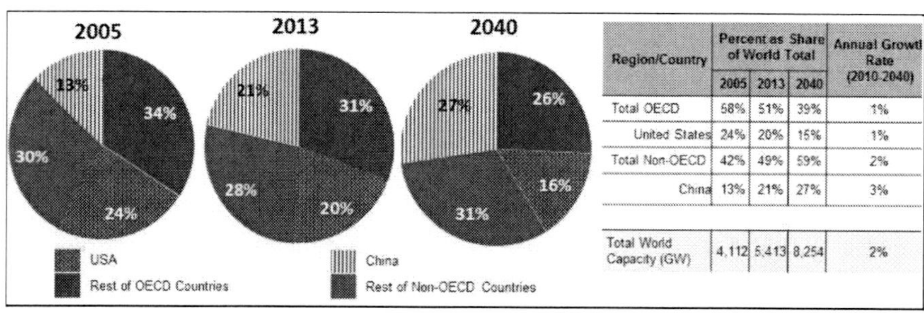

Region/Country	Percent as Share of World Total			Annual Growth Rate (2010-2040)
	2005	2013	2040	
Total OECD	58%	51%	39%	1%
United States	24%	20%	15%	1%
Total Non-OECD	42%	49%	59%	2%
China	13%	21%	27%	3%
Total World Capacity (GW)	4,112	5,413	8,254	2%

Source: "2013 International Energy Outlook," U.S. Energy Information Administration, July 2013.

Figure 9. Global Installed Power Generation Capacity From 2005 to 2040.

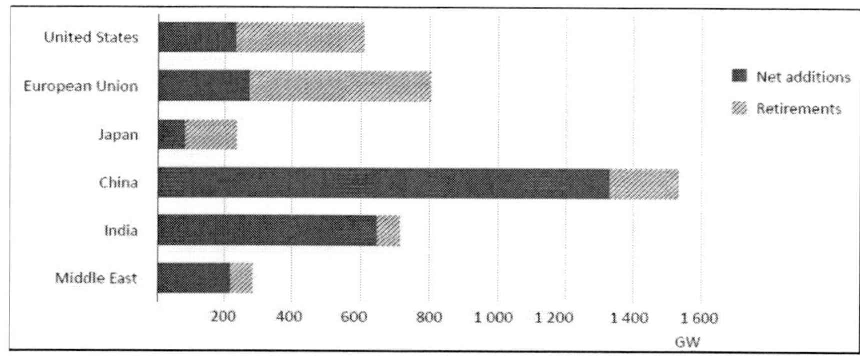

Source: 2013 World Energy Outlook, International Energy Agency, November 2013.

Figure 10. Power Generation Capacity Additions and Retirements between 2013 and 2035.

Similar to EIA, the International Energy Agency's (IEA) 2013 *World Energy Outlook* forecasted that the increase in new power generation capacity in the next two decades would be attributed to non-OECD countries, with China in the lead. IEA projected that China and India together would build almost 40 percent of the world's new capacity between 2013 and 2035, whereas 60 percent of capacity additions in the OECD would replace retired plants (see Figure 10).[57] In other words, while the infrastructure investment needs in developing countries (e.g., China and India) were mainly attributed to new generation capacity additions, the key catalyst for power infrastructure investment in developed countries (e.g., United States) was the replacement market for aging infrastructure. In addition to aging infrastructure, the United States has a need for transmission expansion and upgrade to accommodate new generation connections and maintain electric reliability.

4.2. Transmission Infrastructure Investment in the United States

According to analyses by the Edison Electric Institute (EEI),[58] the U.S. power industry reversed a downward trend in transmission investment in the late 1990s. The uncertainty about the nature and extent of power industry restructuring had triggered a decline in transmission investment in the 1980s and the 1990s. During this period of stagnant investment in transmission infrastructure, the electric load on the Nation's grid more than doubled.[59] This resulted in increasing transmission congestion in certain regions. The long-term trend of declining transmission investment between the 1970s and the 1990s recovered in the late 1990s, and transmission investments grew at a 12 percent annual rate between 1999 and 2003.[60] Reliability and generation interconnection needs were viewed as the main reasons for increasing transmission investments in the United States during this period.

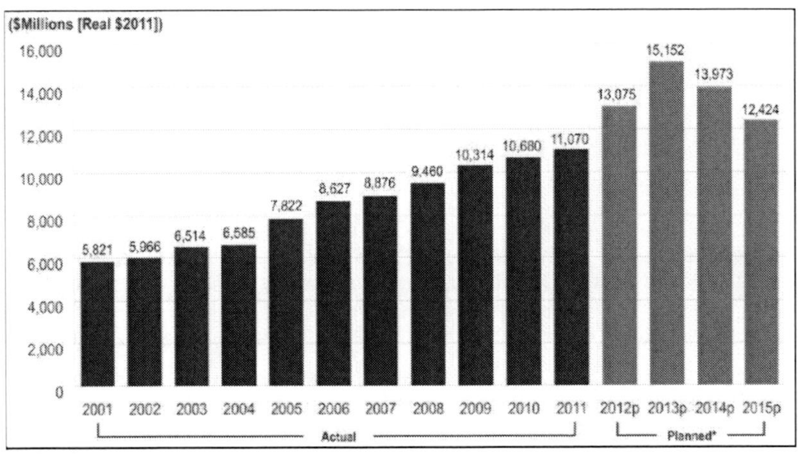

p = preliminary.

Note: The Handy-Whitman Index of Public Utility Construction Costs used to adjust actual investment for inflation from year to year. Forecasted investment data are adjusted for inflation using the GDP Deflator.

* Planned total industry expenditures are preliminary and estimated from the 85percent response rate to EEI's Electric Transmission Capital Budget and Forecast Survey. Actual expenditures from EEI's Annual Property and Plant Capital Investment Survey and from the FERC Form 1 reports.

Source: Edison Electric Institute, Business Information Group.

Figure 11. Estimated Historical and Projected Transmission Investment in the United States From 2001 to 2015.

Figure 11 represents the U.S. transmission investment forecast through 2015, based on (1) EEI's projected capital expenditure growth rates applied to the 2009 U.S. total investment level, and (2) estimated investment requirements associated with transmission circuit-mile additions data from NERC.[61]

NERC data indicated that 22,669 circuit-miles of transmission lines would be added between 2011 and 2015, and that 67 percent of those would be in EHV.[62] On average, the annual transmission investment was forecasted to range from $12 billion to $16 billion between 2013 and 2015. The cost of power transformers accounts for 15 percent to 50 percent of the total transmission capital expenditures; the rest is attributed to the transmission line equipment (e.g., conductors, towers, poles, etc.).[63] Given the developing investment in overall transmission infrastructure, the following section provides a market overview and investment forecast that are specific to power transformers.

5. POWER TRANSFORMER MARKET ASSESSMENT

The global power transformer market is a well-matured one although there are some research and development efforts in high-temperature superconductor transformers and solid state transformers. Analysts have reported that for most of the 1990s, power transformer prices were depressed and that the relationship between sales value and sales volume remained fairly constant.[64] However, starting in 2002, this situation was reversed due to volatile raw commodity prices, unprecedented market demand, and the rationalization of manufacturing bases.[65] Particularly, a sudden rise in the cost of raw materials had a significant impact on the price of power transformers.

According to recent industry analyses, the global power transformer market grew at a compound annual growth rate (CAGR) of about 13 percent from 2000 to 2009, reaching a total revenue of $11 billion in 2009.[66] The global transformer market is forecasted to continue to develop over the next several years. According to a market report released in 2013, the global power transformer market was valued at $17 billion in 2012 and is expected to reach $29 billion by 2019, growing at a CAGR of eight percent from 2013 to 2019.[67] A different source also estimated the global transformer market to grow at a CAGR of eight percent through 2020[68] another estimated that the global electricity transformer market, including distribution transformers, would reach $54 billion (almost 10 million units) by 2017.[69] Key drivers for future transformer market development include an increase in electricity demand in developing countries, replacement of old electric power equipment in matured economies, and a boost for high-voltage power transformers and capital expenditure in the power sector worldwide. In addition, the adoption of energy-efficiency standards in developed markets, such as Europe and the United States, as well as in emerging markets, such as China and India, are expected to create a demand for new, more efficient electricity equipment, including power transformers.[70] The remainder of this section discusses the key suppliers of LPTs, the current condition of the LPT market in the United States, and the domestic manufacturing capacity and historical imports of LPTs. In addition, an overview of China's power transformer industry, including demand, manufacturing capacity, and key manufacturers, is provided in "Appendix E. Power Transformer Industry in China."

5.1. Key Global Suppliers of Power Transformers

The global power transformer market has been one of fluidity and constant adaptation, characterized by a myriad of consolidations, new players, and power shifts throughout the last few decades. Due to the market's shifting nature, the latest statistics for market share by manufacturer are unavailable. Figure 12 provides a list of key power transformer manufacturers in North America in 2012, including new domestic manufacturers.

Large-Power Transformer Suppliers											
	SIEMENS	ABB	HYUNDAI		SMIT	GE	SPX	efacec			
Name	Siemens	ABB	Hyundai	Hico	SMIT	GE - Prolec JV	SPX	EFACEC	CG Pauwels	WEG S.A.	Pennsylvania Transformer
N.A. Plants	Mexico (europe)	Canada Missouri	NEW: Alabama	none (korea)	none (europe)	Mexico	NEW: Wisconsin	NEW: Georgia	Winnepeg	Mexico	Pennsylvania

Source: SPX Transformer Solutions, 2012.

Note: Mitsubishi's new U.S. plant, which was inaugurated in in April 2013 is not included in this figure.

Figure 12. Primary Suppliers of Large Power Transformers in 2012.

In 2011, key manufacturers and analysts reported that ABB, Siemens, and Alstom Grid were the dominant suppliers of power transformers worldwide but indicated that emerging players also had a formidable presence in the global marketplace.[71] According to company reports, ABB had 20 transformer manufacturing plants worldwide in 2011, while Siemens and Alstom Grid had 21 and 13, respectively.[72] In terms of annual production capacity, ABB and Alstom Grid each had approximately 200,000 MVA and 130,000 MVA of annual production capacity, respectively.[73] Alstom Grid's production capacity reflects the additional manufacturing base it obtained through a recent acquisition of AREVA's transmission business in 2010.

5.1.1. Consolidation of Power Equipment Manufacturers

Over the past few decades, the power equipment industry has witnessed a series of mergers and acquisitions (M&A) and consolidation of operations to remove excess capacity and move their operations offshore, fueled by dramatic run-ups in commodity metals prices. In doing so, the firms also took advantage of lower labor costs in certain countries and their rapidly growing domestic electricity demand. Table 5 is a summary of the global M&A activities among power equipment manufacturers that have taken place since the 1980s.

These mergers have reduced the number of firms competing in the global market, and through them, stronger companies have emerged with "increased size, economies of scale, wider product ranges and enhanced financial strength . . . [who benefit] from having greater access to markets and higher bargaining power as a result of combined technological strengths."[74]

5.2. U.S. Power Transformer Market Overview

The United States is one of the world's largest markets for power transformers, with an estimated market value of more than $1 billion in 2010, or almost 20 percent of the global market. The United States also holds the largest installed base of LPTs in the world. Using certain analysis and modeling tools, various sources estimate that the number of EHV LPTs in the United States to be approximately 2,000.[75] While the estimated total number of LPTs (capacity rating of 100 MVA and above) installed in the United States is unavailable, it could be in the range of tens of thousands, including LPTs that are located in medium-voltage transmission lines with a primary voltage rating of 115 kV.

Table 3. Key International Mergers and Acquisitions of Power Equipment Industry From 1980 to 2010

1980s
• GEC (UK) + Alcatel (France) = **GEC Alstom** • ASEA (Sweden) + BBC (Switzerland) = **ABB** • **ABB** (Switzerland): Acquired 39 companies, plus power transmission and distribution (T&D) business of Westinghouse Electric Corporation • to become a technology leader in T&D business
1990s
• In 1998, **Siemens** (Germany) acquired: • Westinghouse's fossil power plant activities • Voith's (Germany) Hydro division • Parson's Power Engineering (UK) • **Babcock Borsig Power** took over B&W (Spain) • **GE Hydro** (Canada) bought Kvaerner (Norway) • GEC Alstom + **ABB** = **ABB-Alstom Power (AAP)**
Early 2000s
• **Alstom** (France) bought ABB's stake in AAP • **GE** (USA) acquired EGT (France) • **Siemens** AG took over Alstom's industrial turbine business • **AREVA** (France) acquired Alstom T&D business • **Hitachi** (Japan) took over the assets of insolvent Babcock Borsig • **JAEPS** (Japan) created a joint venture among Hitachi, Fuji and • Meidensha
2005–2010
• 2005: **Crompton Greaves** (India) acquired Pauwels (Belgium) • 2005: **Siemens** took over VA Tech's (Austria) T&D business • 2008: **ABB** acquired Kulman Electric (US) • 2010: **Alstom** took over AREVA's transmission business   **Schneider** took over AREVA's distribution business

Sources: 1980–early 2000s: http://dhi.nic.in/indian_capital_goods_industry.pdf ;
For 2005–2010:

- Crompton Greaves: http://www.cgglobal.com/pdfs/annual-report/AR_05-06.pdf
- Siemens: http://tdworld.com/news/Siemens-VA-Tech-takeover/
- ABB: http://tdworld.com/business/abb_acquire_kuhlman_electric/
- Alstom/Schneider: http://www.areva-td.com/

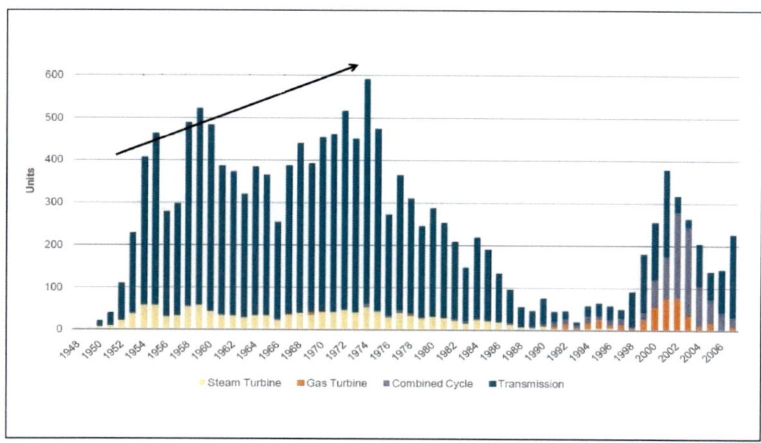

Note: Figure includes LPTs with a capacity rating greater than or equal to 100 MVA
    and excludes replacement demand.
Source: EIA; SPX Electrical Products Group, May 2010.

Figure 13. Yearly Installment of Large Power Transformers in the United States From
1948 to 2006 (Power transformers with a capacity rating greater than or equal to 100
MVA).

Figure 13 represents the historical annual installment of LPTs in the
United States, not including replacement demand. As illustrated in Figure 13, a
large volume of LPTs were installed in the United States between the 1950s
and 1970s. Although investment remained low in the 1990s, the need for LPTs
has been growing steadily since 1999. Despite increasing demand for power
transformers, the United States has a limited domestic capacity to produce
LPTs.

In 2010, the U.S. demand for LPTs was 127,309 MVA, valued at more
than $1 billion.[76] Only 15 percent of the Nation's demand, or 19 percent in
terms of sales value, was met through domestic production. Note, however,
that the data shown in Table 4 includes production capacity and the actual
production of power transformers with a capacity rating of 60 MVA and
above, which is different from the definition of LPT—capacity rating of
greater than or equal to 100 MVA—used throughout this report. Although the
exact figures are unavailable, sources have indicated that the U.S. dependence
on import sources would be much greater than 85 percent for EHV LPTs with
a voltage rating of 345 kV and above. This is due to the limited manufacturing
capacity that existed in the United States prior to 2010, which is further
discussed in the next section.

As a comparison, Figure 14 provides the estimated market size and production of power transformers in the United States and China in 2010. Different parameters are used to define power transformers—a capacity rating of 60 MVA and above for the United States and a voltage rating of 220 kV and above for China.[77] While attributes may vary, the comparison shows the distinct characteristics of the two markets—the U.S.' heavy reliance on foreign-manufactured power transformers and China's abundant domestic production capacity. In 2010, although the estimated market size for the two countries were comparable in the range of 120,000 MVA and 150,000 MVA , the actual production of power transformers in the United States was less than one fifth that of China's.

In terms of manufacturing base in 2010, six domestic manufacturers accounted for all power transformers produced in the United States, whereas more than 30 power transformer manufacturers existed in China. The total annual production capacity[78] of the six domestic factories was approximately 50,000 MVA in 2010, far below the U.S. market demand of 127,309 MVA for that year. On the contrary, China displayed a self-contained LPT market in which the vast majority of its demand was met through domestic production. For example, three of China's largest power transformer manufacturers each held an annual production capacity of more than 100,000 MVA.

Table 4. Summary of the Power Transformer Market (60 MVA +) in the United States in 2010

	Market Share by Quantity		Market Share by Value	
	Quantity (MVA)	Percent (%)	Value (1,000 dollars)	Percent (%)
United States	19,279	15%	$213,070	19%
All Import Sources	108,030	85%	$911,863	81%
Total	127,309	100%	$1,124,933	100%

Note: This analysis includes power transformers with a capacity rating greater than or equal to 60 MVA.

Source: "Large Power Transformers from Korea," USITC, September 2011.

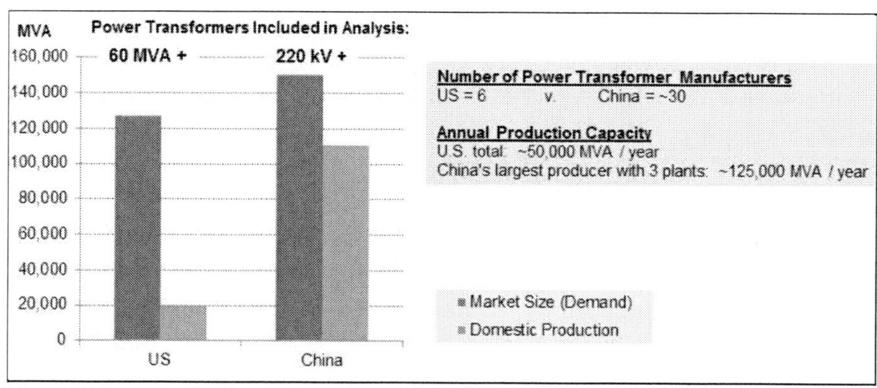

Note: Different criteria used for the United States and China. For the United States, power transformers with a capacity rating greater than or equal to 60 MVA are included in the data; for China, power transformers with a voltage rating greater than or equal to 220 kV are included in the data.

Sources: USITC, 2011; China Transformer Net, 2010; China's Power Transformer Industry Report, Reuters, 2011.

Figure 14. Estimated Power Transformer Markets: United States v. China in 2010.

The following section discusses today's power transformer manufacturing industry in the United States. As a comparison to the global and domestic suppliers, China's manufacturers of LPTs are profiled in "Appendix E. Power Transformer Industry in China."

5.3. Large Power Transformer Manufacturing Capacity in North America

The United States was not an exception to the global, strategic consolidation of manufacturing bases. By the beginning of 2010, there were only six manufacturing facilities in the United States that produced LPTs. Although certain manufacturers reported having the capability to produce power transformers with a capacity rating of 300 MVA or higher, industry experts cautioned that the capacity to produce does not necessarily warrant actual production of power transformers of that magnitude.[79] Often, domestic producers did not have the required machinery and equipment to produce power transformers of 300 MVA, or 345 kV, and above. A number of firms identified constraints in equipment (e.g., cranes, ovens, testing, winding, and vapor phase systems) and the availability of trained personnel set limits to their production capacity.[80]

Up until 2010, a few domestic facilities were capable of producing EHV power transformers although it is unclear whether they actually produced any. While the exact statistics of EHV power transformers produced by domestic facilities are unavailable, sources suggested that the United States procured almost all of its EHV power transformers overseas. However, the limited domestic production capacity has improved significantly since then. Figure 15 is a map of LPT manufacturing facilities in North America in 2013 and the maximum rating of LPTs that they were capable of producing.[81] Also see "Appendix D. Large Power Transformer Manufacturing Facilities in North America" for a detailed list of LPT manufacturing plants in North America.

There were a total of eight power transformer manufacturers in the United States in 2013, four of which were capable of producing LPTs with a voltage rating greater than 345 kV in 2013:

- In April 2010, Efacec's first U.S. transformer manufacturing facility was inaugurated in Rincon, Georgia. Approximately $180 million was invested in this 230,000-square-foot plant which is capable of producing power transformers with ratings up to 1,500 MVA and 525 kV.[82]

- In November 2011, HHI's new 404,000-square-foot power transformer manufacturing plant was inaugurated in Montgomery, Alabama. Since then HHI has been producing large power transformers of capacity ratings up to 550 MVA and 500 kV. With an investment of $100 million, the plant has an annual production capacity of 200 units of 500 kV transformers.[83]

- In April 2013, Mitsubishi completed the construction of a new transformer manufacturing facility in Memphis, Tennessee. Approximately $200 million was invested in this 350,000-square-foot plant which can produce transformers with ratings up to 3,000 MVA and 765 kV and a shipping weight up to 500 tons.[84]

- In April 2012, SPX Transformer Solutions, Inc. (SPX, also formally known as Waukesha Electric Systems, Inc.) completed an expansion at its existing facility in Waukesha, Wisconsin.[85] With a total of 432,00-square-foot and additional crane capacity now up to 500 tons, the plant is expected to produce power transformers with a capacity rating up to 1,000 MVA/500 kV.[86]

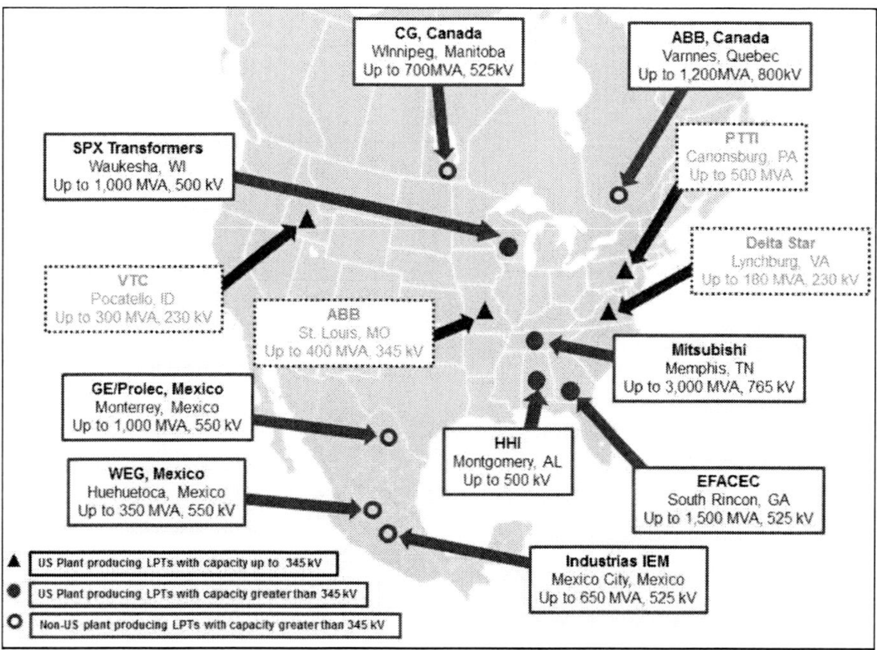

Source: ICF International.

Figure 15. Large Power Transformer Manufacturing Facilities in North America in 2013.

5.4. Historical Imports of Large Power Transformers in the United States

The upward trend of transmission infrastructure investment in the United States which began in 1999 has continued into the 21st century. Figure 16 presents the historical LPT (capacity rating greater than or equal to 100 MVA) imports in terms of value (US$) and quantity (units) between 2005 and 2013.[87] In 2005, the United States imported 363 units, with a total value of $284 million; LPT imports peaked in 2009 with 610 units valued at more than one billion dollars. Since then, LPT imports slowed down a bit. This can be attributed to the fact that four new LPT manufacturing facilities have begun operation in the United States since 2010. Despite newly added domestic capabilities, the demand for LPT imports remained relatively high, with a 496 units totaling $676 million in 2013.[88]

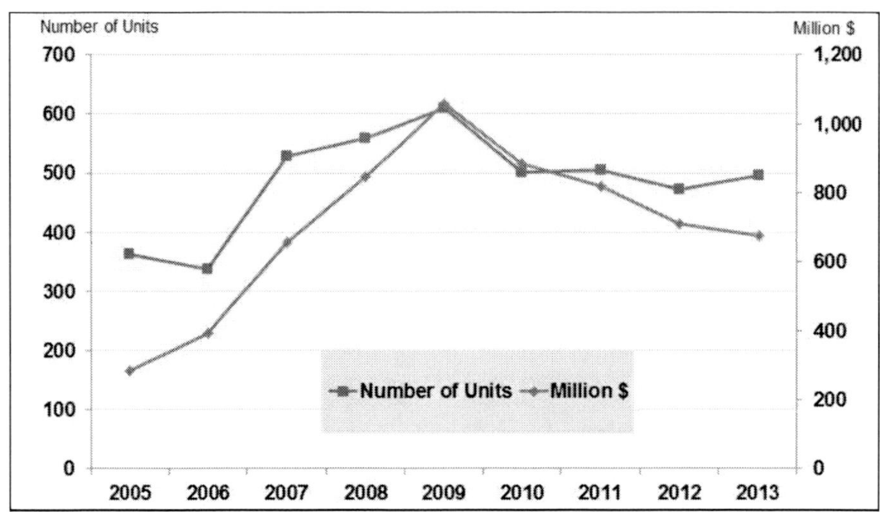

Source: USITC Interactive Tariff and Trade DataWeb, 2014.

Figure 16. U.S. Imports of Large Power Transformers From 2005 to 2013
(Power transformers with a capacity rating greater than or equal to 100 MVA).

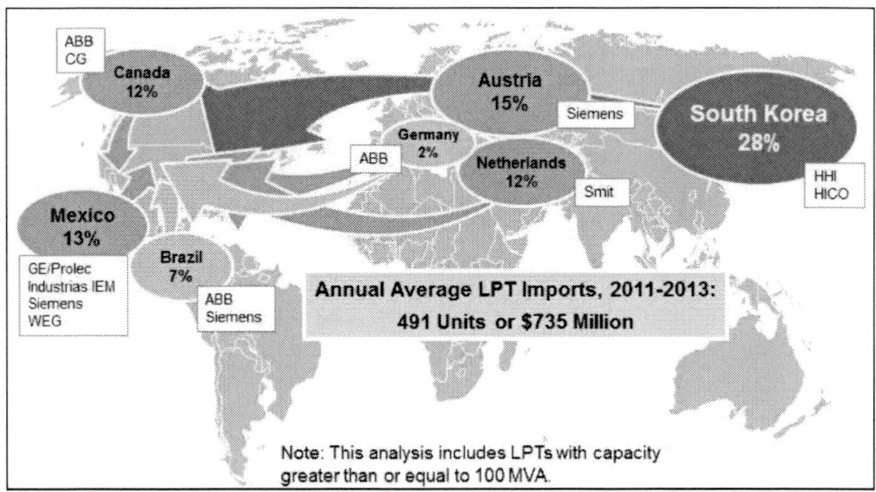

Source: USITC, ICF International Analysis.

Figure 17. Global Suppliers of Large Power Transformers to the United States
Between 2011 and 2013.

Figure 17 is a simplified illustration of the global movement of LPTs to
the United States between 2011 and 2013. Two South Korean firms, HHI and

Hyosung, contributed the largest amount, or 28 percent of total imports in the United States during this three year period. HHI is one of the key global suppliers of LPTs with an annual production capacity of 120,000 MVA at its Ulsan, South Korea plant which is considered one of the largest transformer manufacturing facilities in the world.[89] A few Chinese manufacturers (e.g., TEBA, TWBB, and JSHP) also boast a similar scale of manufacturing capacity (see Appendix E). Other notable global exporters of LPTs to the United States include ABB (from its plants in Canada, Brazil, and Germany), Crompton Greaves of Canada, Siemens of Austria, Prolec GE of Mexico, and Smit Transformer of the Netherlands. A brief profile of these and some of the world's largest power transformer manufacturers is provided in "Appendix F. Selected Global Power Transformer Manufacturers."

As discussed, power transformers are a globally-traded product, and the demand for this equipment is forecasted to continue to grow in the United States. According to two major power transformer manufacturers, the North American power transformer market is expected to grow at a CAGR of three percent to seven percent between 2011 and 2015.[90] In terms of units, one source estimated that the demand for LPTs in the United States would be around 400 to 600 units per year.[91] However, the demand for LPTs has varied greatly over time as the electricity growth rates have changed. In addition to the aging of power transformers, key demand drivers for LPTs include: transmission expansion to integrate solar and wind renewable sources; electric reliability improvement; and new capacity addition in thermal and nuclear power generation.

5.5. Challenges in Global Sourcing of Large Power Transformers

The global demand for LPTs has increased significantly since the late 1990s to meet the power demand growth and the need to replace aging power transformers. Purchasing decisions in the global marketplace are more complex than simply comparing prices. Some of the challenges associated with global procurement of LPTs are as follows:[92]

- Ocean and inland transportation, compliance with specifications, quality, testing, raw materials, and major global events (e.g., hurricanes) can significantly influence a supplier's lead time and delivery reliability. In addition, some railroad companies are removing rail lines due to infrequent use and other lines are not being maintained. This can pose a challenge to moving the LPTs to certain locations where they are needed.

- Fluctuations in currency exchange rates and the prices of materials during the time in which a power transformer is being manufactured can quickly change the competitive bid price for the order.
- Cultural differences and other communication barriers can be challenging. In many cultures, what the buyer-manufacturer relationship entails may vary from what is written in the contract.
- Foreign factories may not understand the U.S. standards such as the Institute of Electrical and Electronics Engineers (IEEE) and the National Institute of Standards and Technology (NIST) or have appropriate testing facilities.
- Foreign vendors may not have the ability to repair damaged power transformers in the
- United States.
- It is expensive to travel overseas for quality inspections and to witness factory acceptance testing.
- The utility industry is also facing the challenge of maintaining an experienced, well- trained in-house workforce that is able to address power transformer procurement and maintenance issues.

Utilities can minimize the potential risks related to global sourcing by focusing on proactive business strategies, planning effectively, and managing a portfolio of qualified and experienced suppliers.

6. RISKS TO POWER TRANSFORMERS

As discussed in this report, LPTs are significant investment pieces that are critical to the reliable operation of the electric grid; therefore, the assessment of the health of and risks to LPTs is an essential part of proper maintenance of the equipment. Figure 18 is an analysis of the main causes of power transformer failures between 1991 and 2010. This figure is based on the examination of historical insurance claims for various utility type transformers during the 20-year period, which included several hundred transformer failures.[93] This assessment was based on an insurance firm's own internal investigation of the failures. As shown in Figure 18, electrical disturbances were the leading cause for power transformer failures, responsible for 28 percent of the total failures that occurred during this 20-year period.[94] "Electrical disturbances" included phenomena such as switching surges, voltage spikes, line faults/flashovers, and other utility abnormalities, but excludes lightning.

Although age is not included as a cause of transformer failure in Figure 18, age is certainly a contributing factor to increases in transformer failures. Various sources, including power equipment manufacturers, estimated that the average age of LPTs installed in the United States is 38 to 40 years, with approximately 70 percent of LPTs being 25 years or older.[95] According to an industry source, there are some units well over 40 years old and some as old as over 70+ years that are still operating in the grid. An LPT is subjected to faults that result in high radial and compressive forces, as the load and operating stress increase with system growth.[96] In an aging power transformer failure, typically the conductor insulation is weakened to the degree at which it can no longer sustain the mechanical stresses of a fault.[97]

Given the technical valuation that a power transformer's risk of failure is likely to increase with age, many of the LPTs in the United States are potentially subject to a higher risk of failure. Although age can be a factor, the life expectancy of a power transformer varies depending on how it is used. In addition, according to an industry source, there were also some bad batches of LPTs from certain vendors. The same source also estimated that the failure rate of LPTs is around 0.5 percent. In addition to these traditional threats to power transformers, the physical security of transformers at substations has become a public safety concern due to a coordinated physical attack on cyber infrastructure of a California substation in 2013. Efforts are under way to increase utilities' awareness of possible substation vulnerabilities.

In recognition of the importance of LPTs with regard to the reliability of the grid, there are various ongoing efforts to enhance the resilience of power transformers. Specifically, there is an increasing amount of activities to address the potential threats to LPTs, including the following:

- On March 7 2014, the Federal Energy Regulatory Commission (FERC) directed NERC to develop mandatory physical security standards within 90 days in the wake of attacks on transmission facilities in the United States in 2013. Owners and operators are to first identify critical facilities, and then develop and implement plans to protect against physical attacks that may compromise the operability or recovery of such facilities.[98]

- NERC, under the direction of FERC, are developing reliability standards that are intended to mitigate the effects of GMDs on the reliable operation of the electric power system, including power transformers in a two-stage effort. The phase-one reliability standard will require applicable registered entities to develop and implement operating procedures that can mitigate the effects of GMD events.[99] Phase two efforts will require applicable registered entities to conduct initial and on-going assessments of the potential impact of benchmark GMD events on their respective system.[100]

- A number of manufacturers are exploring the development and implementation of mitigation and hardening options, including the development of parts that are more resilient to potential threats, as well as protective devices.

- EEI's Spare Transformer Equipment Program and NERC's Spare Equipment Database Program are designed to provide ways in which utilities may identify and share spare transformers across North America during an emergency.[101] As this information becomes available, this will help decision makers understand what additional programs or incentives may be needed to increase the number of available spares.

- The U.S. Department of Homeland Security's (DHS) Science and Technology Directorate, along with their partners, the Electric Power Research Institute, ABB, and CenterPoint Energy (CNP), and with the support of DOE and the DHS Office of Infrastructure Protection, have developed the Recovery Transformer (RecX), a prototype EHV transformer that would drastically reduce the recovery time associated with EHV transformers. The RecX is lighter (approximately 125 tons), smaller, and easier to transport and quicker to install than a traditional EHV transformer. The RecX has been operating in CNP's grid since March 2012, after a successful exercise that included the transportation, installation, assembly, commissioning and energization of the transformer in less than one week. The RecX is a 345:138kV, 200 MVA per phase transformer (equivalent to 600 MVA) and was designed to be an applicable replacement for more than 90 percent of transformers in this voltage class, which is the largest voltage class of EHV transformers.[102]

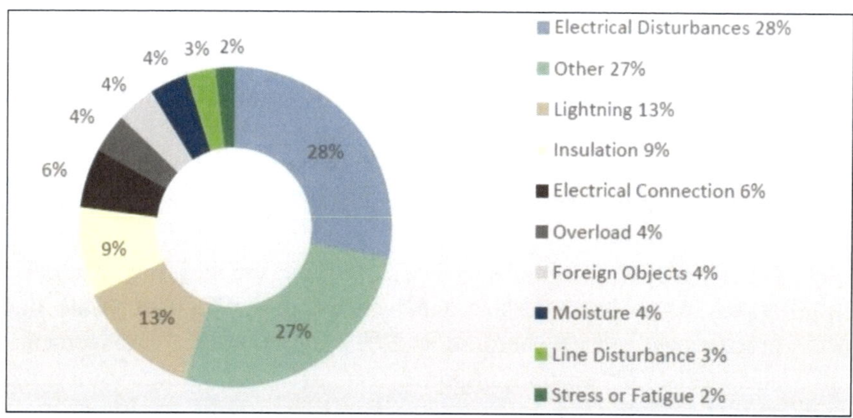

Source: SPX; 2012 Doble Engineering Company—79th Annual International Doble Client Conference; Analysis of Transformer Failures, by William H. Bartley P.E., Hartford Steam Boiler Inspection & Insurance Co.

Figure 18. Causes of Transformer Failures Between 1991 and 2010 (as a Percentage of Total Failures).

CONCLUDING REMARKS

This report examined the global power-transformer manufacturing industry, one that is characterized by continuous adaptation due to shifting and unpredictable market dynamics. In particular, this study addressed the considerable dependence the United States has on foreign suppliers to meet its growing need for LPTs. The intent of this study is to inform decision- makers about potential supply concerns regarding LPTs in the United States. This report provides the following observations:

- The demand for LPTs is expected to remain strong globally and domestically. Key drivers of demand include the development of power and transmission infrastructure in emerging economies (e.g., China and India) and the replacement market for aging infrastructure in mature economies (e.g., United States), as well as the integration of alternative energy sources into the grid and an increased focus on nuclear energy in light of climate change concerns.

- Two key raw materials—copper and electrical steel—are vital to LPT manufacturing. However, there are limited supply sources for the special grades of electrical steel needed for the LPT core, and both steel and copper have experienced wide price fluctuations since 2004. Although the prices of electrical steel have generally declined from their 2008 highs, the sourcing of key raw materials remain essential to LPT procurement and manufacture.

- LPTs require a long lead time, and transporting them can be challenging. The average lead time for an LPT is between five and 16 months; however, the lead time can extend beyond 20 months if there are any supply disruptions or delays with the supplies, raw materials, or key parts. Its large size and weight can further complicate the procurement process, because an LPT requires special arrangements and special rail cars for transport.

- The United States has limited production capability to manufacture LPTs. In 2010, only 15 percent of the Nation's demand for power transformers (with a capacity rating of 60 MVA and above) was met through domestic production. Although the exact statistics are unavailable, power transformer market supply conditions indicate that the Nation's reliance on foreign manufacturers was even greater for EHV power transformers with a capacity rating of 300 MVA and above (or a voltage rating of 345 kV and above).

- However, domestic production of LPTs is expected to continue to improve. Since 2010, four new or expanded facilities have begun producing EHV LPTs in the United States: Efacec in Rincon, Georgia; HHI in Montgomery, Alabama; Mitsubishi in Memphis, Tennessee; and SPX's expanded plant in Waukesha, Wisconsin. While these plants may not have reached their full capacity yet, it is evident that the new domestic capacities will provide some relief with regard to the U.S. dependence on foreign manufacturers for LPTs.

The U.S. electric power grid serves one of the Nation's critical life-line functions on which many other critical infrastructure depend, and the destruction of this infrastructure can cause a significant impact on national security and the U.S. economy. The U.S. electric power grid faces a wide variety of threats, including natural, physical, cyber, and space weather. LPTs are large, custom-built electric infrastructure. If several LPTs were to fail at the same time, it could be challenging to quickly replace them. While the potential effects of these threats on the electric power grid are uncertain, the Electricity Sector continues to work on a variety of risk management strategies to address these potential severe risk impacts. Understanding the characteristics of today's power transformer procurement and supply environment is indispensable to both the public and private sectors. The assessment of LPTs in this DOE report provides background to industry and government stakeholders as they continue their efforts to enhance critical energy infrastructure resilience in today's complex, interdependent global economy.

APPENDIX A. ACRONYMS AND ABBREVIATIONS

CAGR	Compound annual growth rate
CG	Crompton Greaves
CNP	CenterPoint Energy
CRGO	Cold-rolled grain-oriented
DHS	U.S. Department of Homeland Security
DOE	U.S. Department of Energy
EEI	Edison Electric Institute
EHV	Extra high voltage
EIA	U.S. Energy Information Administration
FERC	Federal Energy Regulatory Commission FOB Free on Board
GMD	Geomagnetic disturbance
HHI	Hyundai Heavy Industries
HICO	Hyosung Power and Industrial Systems
HILF	High-Impact Low-Frequency
HSPD-7	Homeland Security Presidential Directive-7
HV	High-voltage
IEA	International Energy Agency
IEEE	Institute of Electrical and Electronics Engineers
ISO	International Organization for Standardization
JSHP	JiangSu HuaPeng Transformer Co., Ltd.
Kg	Kilograms
kV	Kilovolts
LPT	Large power transformer M&A Mergers and acquisitions
MVA	Megavolt-amperes
NERC	North American Electric Reliability Corporation
NIPP	National Infrastructure Protection Plan
NIST	National Institute of Standards and Technology
OECD	Organization for Economic Cooperation and Development
PTTI	Pennsylvania Transformer Technology, Inc.
ReX	Recovery Transformer

SGCC	State Grid Corporation of China
TBEA	Tebian Electric Apparatus Stock Co., Ltd.
TWBB	Baoding Tianwei Baobian Electric Co., Ltd.
USITC	United States International Trade Commission
VTC	Virginia Transformer Corp.
W/kg	Watts per kilogram

APPENDIX B. ELECTRICAL STEEL MANUFACTURER PROFILES

Name	Description
AK Steel	AK Steel, founded in 1899 and headquartered in Middletown, Ohio, employs about 6,600 people in Ohio, Kentucky, Indiana, and Pennsylvania. With $6 billion in sales in 2010, the company produces flat-rolled carbon, stainless, and electrical steel products. AK Steel produces a range of electrical steels, including oriented steel grades of M2, M3, M4, M5, and M6, non-oriented standard steel grades of M15 to M47, and domain-refined, laser scribed steels, H-0 DR, H-1 DR, and H-2 DR. In 2009, AK Steel produced 312,000 tonnes of grain oriented electrical steel.
Allegheny Ludlum	Allegheny Ludlum Corporation, headquartered in Pittsburgh, PA operates specialty metals manufacturing facilities in Pennsylvania, Connecticut, Massachusetts, Indiana, and Ohio. In 2009, Allegheny Ludlum had revenue of approximately $3.1 billion. In addition to its other stainless and specialty steel products, it produces grain oriented steel with grades from M2 to M6. In 2009, Allegheny Ludlum produced 109,000 tonnes of grain oriented electrical steel
Angang Steel	Angang Steel, located in China, was incorporated in 1997 with Anshan Iron and Steel Group Complex as its sole promoter. It produces a wide array of steel products, and began the mass production of grain oriented steel at the beginning of 2011. The facilities are expected to have the capacity to produce 100,000 tonnes of grain oriented electrical steel annually.
ArcelorMittal	ArcelorMittal, a Brazilian company, was founded in 1944 and produced a total of 90.6 million tonnes of crude steel in 2010, representing approximately 8 percent of world steel output. ArcelorMittal offers both grain oriented and non-oriented electrical steel. In 2010, ArcelorMittal proposed a 50-50 joint venture with SAIL to set up a steel facility in Bokaro, India.

Name	Description
	Additionally, in 2008 ArcelorMittal entered into a joint venture with Hunan Valin Steel Group to build a steel facility in China with a projected annual capacity of 200,000 tonnes of grain oriented steel. In 2009, ArcelorMittal produced 107,000 tonnes of grain oriented electrical steel.
China Steel Corporation	China Steel Corporation, the only integrated steel producer in Taiwan, was founded in 1971 and exports approximately 25 percent of its steel production volume. It currently has an annual crude steel production capacity of approximately 13.4 million tonnes and produces four grades of electrical steel. China Steel Corporation does not produce grain oriented electrical steel. CSC is planning to invest $486 million in a new production line for electrical steel which would produce 150,000 to 200,000 tonnes of electrical steel annually.
Duferco Viz Stal Metallurgic al Plant	The Viz Stal plant was founded in 1726, as a pig iron processing facility. In 1914, the plant became the first producer of hot-rolled, nonoriented steel in Russia. Then, in 1934 it began producing hot-rolled, grain-oriented steel. In 1973, the plant began producing cold-rolled, grain-oriented electrical steel, and in 1978 Viz Stal became the first manufacturer of cold- rolled, nonoriented electrical steel in the Soviet Union. Duferco, a Swiss international manufacturing and trading company, acquired the plant. Finally, in 2004, Viz Stal gained the capability to supply their customers with slit coils.
JFE Steel Corporation	JFE Steel Corporation is a Japanese company formed in December 2001 from a merger between Kawasaki Steel and NKK Corporation. It produced a total of 29.3 million tonnes of crude steel in 2008. JFE Steel produces several types (each with several grades) of grain oriented electrical steel and non-oriented electrical steel. In 2009, JFE Steel Corporation produced 160,000 tonnes of grain oriented electrical steel.
Nippon Steel Corporation	In 1970, Yawata Iron and Steel and Fuji Steel merged to form Nippon Steel Corporation. Located in Tokyo, Japan, Nippon Steel has about 15,800 employees and produces almost 30 million metric tons of crude steel annually. Nippon produces grain oriented steel with grades from M2 to M6 and non-oriented standard steel grades of M15 to M45. In 2009, Nippon Steel produced 243,000 tonnes of grain oriented electrical steel. In 2009, 20 percent of Nippon Steel's exports went to China, while 60 percent were distributed among other Asian regions, 5 percent to North America, 4 percent to South America, and 2

Appendix B. (Continued)

Name	Description
	percent to Europe. The remaining 9 percent were disbursed between Africa, the Middle East, and Oceania
Novolipetsk (NLMK) Metallurgic al Plant	NLMK started in 1931 when iron ore and limestone deposits were discovered in Lipetsk, Russia. NLMK is now the largest steel sheet producer in Russia, and produced more than 11 million metric tons of crude steel in 2010. In 2006 it acquired Viz Stal Metallurgical Plant, which was the largest producer of grain oriented steel and the second largest electrical steel producer in Russia. NLMK's share of global grain oriented electrical steel production is nearly 11 percent and more than 80 percent of its products are exported. In 2007, it produced 189,000 tonnes of grain oriented steel and 19,000 tonnes of non-grain oriented steel. In 2009, NLMK produced 344,000 tonnes of grain oriented electrical steel. NLMK's total transformer steel production capacity is approximately 350,000 tonnes annually.
Pohang Iron and Steel (POSCO)	POSCO, located in the port city of Pohang, South Korea, was founded in 1958, produced 35 million tonnes of steel in 2010, and has approximately 30,000 employees. In 2009, POSCO had about one million tonnes of electrical steel capacity annually. POSCO is considering partnering with Steel Authority of India Limited (SAIL) to build a production complex with a proposed annual capacity of about 3 million tonnes. In 2009, POSCO produced 250,000 tonnes of grain oriented electrical steel.
Shanghai Baosteel	Shanghai Baosteel, formerly Baoshan Iron & Steel, is state owned and China's largest iron and steel maker. Baosteel and its 22 wholly-owned subsidiaries have an annual production capacity of around 30 million metric tons of crude steel and 600,000 tonnes of electrical steel. In 2009, it produced 90,000 tonnes of grain oriented electrical steel.
Stalprodukt S.A.	In 1992, the Polish company Stalprodukt S.A. purchased two former Sendzimir Steel Works production plants. Stalprodukt S.A. produces four grades of grain oriented electrical steels. In 2009, Stalprodukt S.A. produced 62,000 tonnes of grain oriented electrical steel.
Tata Steel (Cogent Power Ltd.)	Tata Steel is the world's tenth largest steel producer, with a crude steel capacity of more than 28 million tonnes. In April 2007, Tata Steel acquired Corus, an international metal company that provides electrical steel through its wholly-

Name	Description
	owned subsidiary, Cogent Power Ltd. The electrical steel division of Cogent Power comprises Orb Works, located in South Wales, and Surahammars Bruk, headquartered in Sweden. Orb Works and Surahammars both produce a wide variety of both grain oriented and non-oriented steels. In 2009, Orb produced 90,000 tonnes of grain oriented electrical steel.
ThyssenKr upp Steel	ThyssenKrupp Steel, a subsidiary of ThyssenKrupp AG, entered the electrical steel market in 1989. In 2002 ThyssenKrupp Electrical Steel (TKES) was formed to consolidate all of the company's electrical steel activities. Further restructuring in 2004 created ThyssenKrupp Stahl AG to handle the company's non-oriented electrical steel products. TKES now deals solely with grain oriented steels. TKES is headquartered in Essen, Germany and has plants in Germany, India, Deutschland, Italy and France. ThyssenKrupp has a production capacity of approximately 1.5 million tonnes of electrical steel annually, making it the largest electrical steel producer in Europe and the third largest producer worldwide. In 2009, TKES produced 250,000 tonnes of grain oriented electrical steel.
WCI Steel	WCI, a U.S. manufacturer based in Warren, Ohio, produced nonoriented electrical steel until exiting the business in January 2004. In September 2003, WCI filed for protection under Chapter 11 of the U.S. Bankruptcy Code. WCI management cited continuing volume deterioration and negative profit margins for the halt of steel production.
Wuhan Iron and Steel (WISCO)	WISCO, a Chinese company, produced 30 million metric tons of crude steel in 2009, and increased its annual electrical steel capacity to 2 million metric tons in 2010 with the completion of three new production lines. Currently, more than half of domestic silicon steel demand in China is met by WISCO. In 2009, WISCO produced 440,000 tonnes of grain oriented electrical steel, making it the largest producer of grain oriented steel.

Source: Distribution Transformers Final Rule Technical Support Document, Appendix 3A. Core Steel Market Analysis, DOE, Office of Energy Efficiency and Renewable Energy, April 2013, http://www1.eere.energy.gov/buildings/ appliance_standards/commercial/pdfs/transformer_preanalysis_app3a.pdf (accessed January 14, 2014).

APPENDIX C. HISTORICAL IMPORTS OF LARGE POWER TRANSFORMERS IN THE UNITED STATES

Table C1. Large Power Transformer Imports by Country From 2005 to 2013 (in US $ value)
(Liquid dielectric transformers having a power handling capacity exceeding 100,000 kVA)

Country	2005	2006	2007	2008	2009	2010	2011	2012	2013	Market Share 2011-2013 (%)
	In 1,000,000 Dollars									
Korea	33	73	173	255	292	331	279	210	126	28%
Canada	60	56	63	88	132	133	85	98	85	12%
Austria	46	55	84	102	132	106	109	126	93	15%
Mexico	51	94	171	159	165	91	91	97	101	13%
Netherlands	27	31	35	63	48	67	88	61	100	11%
Brazil	7	23	47	51	87	30	23	51	45	5%
Germany	16	30	23	47	54	11	26	28	13	3%
Subtotal	240	362	596	765	910	769	701	671	563	88%
All Other	44	30	63	81	147	114	116	40	113	12%
Total	**284**	**391**	**658**	**847**	**1,057**	**883**	**817**	**711**	**676**	**100%**

Table C2. Large Power Transformer Imports by Country from 2005 to 2013 (by quantity)
(Liquid dielectric transformers having a power-handling capacity exceeding 100,000 kVA)

Country	2005	2006	2007	2008	2009	2010	2011	2012	2013	Market Share 2011-2013 (%)
	In Actual Units of Quantity									
Korea	82	77	123	160	180	223	204	142	76	29%
Mexico	57	85	190	156	107	63	77	97	162	23%
Canada	56	36	34	34	44	43	26	37	28	6%

Country	2005	2006	2007	2008	2009	2010	2011	2012	2013	Market Share 2011-2013 (%)
					In Actual Units of Quantity					
Austria	64	38	47	45	69	33	43	76	59	12%
Netherlands	24	21	21	33	22	30	40	36	72	10%
Brazil	9	17	27	25	37	12	13	30	27	5%
Germany	11	21	23	42	36	11	26	16	9	3%
Subtotal	303	295	465	495	495	415	429	434	433	88%
All Others	60	42	63	64	115	86	76	38	63	12%
Total	**363**	**337**	**528**	**559**	**610**	**501**	**505**	**472**	**496**	**100%**

Source: USITC Interactive Tariff and Trade DataWeb, http://dataweb.usitc.gov/scripts/
user_set.asp, accessed January 5, 2014.

APPENDIX D. LARGE POWER TRANSFORMER MANUFACTURING FACILITIES IN NORTH AMERICA

Firm	Plant Location	Types of Transformers Produced at Location	Notes
ABB	South Boston, VA St. Louis, MO	Small to large size (up to 400 MVA and 345 kV) and midsize	
Delta Star	Lynchburg, VA San Carlos, CA	Small to medium size of 5-230 kV and 5-180 MVA	
Efacec	South Rincon, GA	Both core- and shell-type transformers of up to 1,500 MVA and 525 kV	Inaugurated in April 2010
PTTI	Canonsburg, PA	From 60 MVA up to 500 MVA	
SPX (Waukesha)	Goldsboro, NC Waukesha, WI	Mid to large size power transformers of up to 1,000 MVA and 500 kV Had limited capacity to produce	Expansion completed in April 2012

Appendix D. (Continued)

Firm	Plant Location	Types of Transformers Produced at Location	Notes
		EHV prior to 2012 expansion	
VTC	Roanoke, VA Pocatello, ID	Up to 300 MVA / 230 kV	
HHI	Montgomery, AL	Up to 500 kV (200 units/year once fully operational)	Inaugurated in November 2011
Mitsubishi	Memphis, TN	N/A	Inaugurated in April 2013
ABB	Varnnes, Quebec	Up to 1,200 MVA / 800 kV	Outside the USA
Crompton Greaves	Winnipeg, Manitoba	Up to 700 MVA / 525 kV	Outside the USA
Industrias IEM	Mexico City, Mexico	Up to 650 MVA / 525 kV	Outside the USA
Prolec GE	Monterrey, Mexico	Up to 1,000 MVA / 550 kV	Outside the USA
WEG	Huehuetoca, Mexico	Up to 350 MVA / 550 kV	Outside the USA

Source: Open source research.

APPENDIX E. POWER TRANSFORMER INDUSTRY IN CHINA

China is the largest rising market for power transformers in both manufacturing capacity and demand, and whose market conditions undoubtedly affect those of the world. The vast majority of China's power transmission system is run by the State Grid Corporation of China (SGCC), a state-owned enterprise that constructs, owns, and operates the transmission and distribution systems in China's 26 provinces.[103] Established in 2002, the SGCC implemented a centralized bidding system for the procurement of transmission and distribution equipment.

China's transformer market is dominated by 220 kV transformers and below, despite the strengthening of super-high and ultra-high voltage power grid construction in recent years. In 2011, approximately 81 percent of the

SGCC's bidding capacity was for 220 kV and below, while power transformers of 500 kV and above occupied less than 17 percent.[104] In 2010, the total bidding capacity for power transformers of 220 kV and above was 150,337 MVA, after a 25 percent decline from the prior year.[105] At present, China's transformer market shows signs of oversupply, boasting approximately 30 manufacturers of transformers of 220 kV and above and more than 1,000 manufacturers of power transformers of 110 kV and below.[106]

In addition to local companies, large multinational firms—Siemens and ABB in particular—also play main roles in the manufacturing of 220 kV and above transformers in China. With manufacturing facilities in the mainland, they shared 20 to 30 percent of China's transformer market in 2011 and continue to expand their footprint by leveraging advanced technology.[107] As of December 2011, Siemens had three transformer manufacturing plants in China, boasting a total annual production capacity of over 70,000 MVA;[108] ABB had five transformer manufacturing facilities in China, and its largest facility in Chongqing had an annual production capacity of 50,000 MVA.

Table E1 provides the top eight Chinese transformer manufacturers in terms of sales value in 2009. As of 2011, the primary market for these Chinese manufacturers was inland; however, some of the firms have begun extending their footprints overseas. With what appears to be large excess capacity, some Chinese manufacturers were already exporting transformers abroad. Some of them have established sales offices in the United States, including TBEA, TWBB, and JSHP.

Table E1. Top Chinese Power Transformer Manufacturers in 2009

Company Name	Annual Production Capacity*	Notes	Contact in the United States	Web site
Tebian Electric Apparatus Stock Co., Ltd (TBEA)	100,000 MVA (Two plants combined)	China's largest transformer manufacturer produces transformers of up to 1,000 kV, with a technical advantage on transformers 220 kV and above.	3452 E. Foothill Blvd., Ste. 1020 Pasadena, CA Phone: 001–626–7921037 Fax: 001–626–6283459 Email: shilin@tbea-usa.com	http://www.tbea-usa.com/ http://en.tbea.com.cn/
JiangSu HuaPeng Transformer Co., Ltd.	100,000 MVA	In 2008, JSHP delivered 832 transformer units with a capacity of 110 kV–345 kV,	4030 Moorpark Ave., Ste. 222 San Jose, CA 95117 Phone: +1–408–850–1416	http://www.jshp.com/

Table E1. (Continued)

Company Name	Annual Production Capacity*	Notes	Contact in the United States	Web site
(JSHP Transformer)		or a total of 59,253 MVA.	Fax: +1–408–519–7091 Email: sales@jshp.com	
Xi'an XD Transformer Co., Ltd (XD Transformer)	50,000 MVA	Produces a wide range of transformers (10 kV–1000 kV) and has a technical advantage with transformers of 110 kV and above.	N/A	http://www.xdxb.com.cn/English/index.asp
Baoding Tianwei Baobian Electric Co., Ltd. (TWBB)	125,000 MVA (Three plants combined)	Exports 230 kV / 400 MVA transformers to more than 30 countries, including the United States.	Rotterdam, New York 12303 Phone: 518–357–9290 Cell: 518–421–4081 Email: wpesales@worldpowerequipment.com	http://www.twbb.com/web/einfo.asp?bid=8
Shandong DaChi Electric	N/A	Main products include 550 kV/400 MVA power transformers, 35kV/200 MVA, and distribution transformers.	N/A	http://www.chinadachi.com/index.asp
Sunten Areva Electric	N/A	Mainly produces 35 kV distribution transformers.	N/A	http://www.sunten.com.cn/ENGLISH_B/index.asp
Changzhou Xiandian Transformer	28,000 MVA	Produces transformers of a voltage rating of 500 kV and below.	N/A	http://www.czxd.com.cn/en/about.asp
Hangzhou Qianjiang Electric Group	N/A	Produces transformers with a voltage rating of up to 400 kV and distributes transformers.	N/A	http://www.qre.com.cn/en/index.aspx

* Note: The term Annual Production Capacity (MVA) represents a combined capability of a manufacturer to produce a range of transformers (from small to large) from all of its facilities in an annual basis.

Source: "2009 - Top 10 China Transformer Manufacturers," China National Transformer Association, June 22, 2010, http://www.dsius.com/ Top10_TR_in China.pdf (accessed March 24, 2014).

APPENDIX F. SELECTED GLOBAL POWER TRANSFORMER MANUFACTURERS

Company Name	Number and Location of Plants	Annual Production Capacity (MVA)*	LPT Production	Presence and Primary Markets Served	Notes
ABB	20 worldwide (Brazil, Canada, China, Germany, India, Poland, Spain, Sweden, Thailand, Turkey, USA)	200,000 MVA (50,000 MVA in Chongqing, China; 15,500 MVA in Quebec, Canada)	Full-range LPTs up to 800kV/1200 MVA	Worldwide	Remains the largest transformer manufacturer and the global leader in transformer technology
Alstom (AREVA)	13 worldwide	130,000 MVA	Full-range LPTs up to 1200 kV	Worldwide	Primary markets are Asia/Pacific, Americas, and Europe/ Africa
Bharat Heavy Electricals	India	45,000 MVA		Asia, Middle East, Africa, Europe	
Crompton Greaves Ltd.	(Belgium, Canada, Hungary, India, Indonesia, Ireland, France, UK, USA)	70,000 MVA	100 - 1100 MVA		Acquired Belgium-based Pauwels Group and gained facilities in 5 countries: Belgium, Ireland, Canada, USA, and Indonesia
HHI	3 (South Korea,	120,000 MVA (Excluding	Full range LPTs up to	Worldwide; Market presence in	Construction of a new U.S. plant

Appendix F. (Continued)

Company Name	Number and Location of Plants	Annual Production Capacity (MVA)*	LPT Production	Presence and Primary Markets Served	Notes
	Bulgaria, USA)	a new U.S. plant)	800 kV/1500 MVA	2009: North America (46%) Middle East (28%)	completed in November 2011
HICO	4 (South Korea, China)	75,000 MVA	765 kV / 2200 MVA	Korea; North America	
Mitsubishi	Japan	N/A	1050 kV / 3000 MVA	Worldwide; North America	
Prolec GE	Mexico	100,000 MVA	Full-range LPTs up to 550 kV / 1000 MVA	Americas, Africa, the Middle East	
Siemens (VA Tech)	21 worldwide (China, India, Austria, Brazil)	Total production capacity unavailable. 70,000 MVA in China	Full-range LPTs up to 800 kV / 1000 MVA	Worldwide	Primary markets are Europe, China, and North America
Smit Transformers	Netherlands, Germany, Malaysia	N/A	800 kV / 1200 MVA	Germany, Netherlands, USA, Malaysia	Exports to the USA
SPX	USA	N/A	500 kV / 1200 MVA	North America	Expansion completed in April 2012 in Waukesha, Wisconsin

*Note: The term Annual Production Capacity (MVA) represents a combined capability of each manufacturer to produce a range of transformers (from small to large) from all of its facilities on an annual basis.

APPENDIX G. BIBLIOGRAPHY

ABB Group. "ABB unveils its largest transformer components center in China." Retrieved April 1, 2014 from http://www.abb.mu/cawp/seitp202/51719ef8b8b364ee4825769 0001f063a.aspx

ABB Group. (2002, November 26). "ABB signs US$ 200 million agreement for power transformers in United States." Retrieved April 1, 2014 from http://www.abb.com/cawp/seitp202/c1256c290031524bc1256c7d003d799 2.aspx

ABB Group. "The evolution of power transformers: 110 years of power transformer technology." Retrieved December 7, 2011 from http://www.abb.com/cawp/db0003db002698/1d5674d769397bc8c12572f4 0045bd75.aspx

ABB Group. "Railcar rental program for power transformer relocation." Retrieved April 1, 2014 from http://www05.abb.com/global/scot/ scot252.nsf/veritydisplay/36bcc4e173d5c1558525760b00711641/$file/1z ul004605-300_railcar_r4.pdf

ABB Group. (2007). "Service handbook for transformers." ABB Group. (2007, September 5). ABB Strategy 2011. Alstom Grid. (2011, June 28). Shareholders Meeting.

Alstom Grid. "Queensland Minister commemorates successful completion of upgrade project at Alstom Grid's Rocklea factory." Retrieved April 1, 2014 from http://www.alstom.com/grid/news-and-events/press-releases/upgrade-project-at-Alstom- Grid-s-Rocklea-factory/

Alstom Grid. "Power Transformers." Retrieved April 1, 2014 from http://www.alstom.com/grid/solutions/high-voltage-power-products/ electrical-power- transformers/

Bartley, William H., "Analysis of Transformer Failures," Hartfort Steam Boiler Inspection & Insurance Co., 79th International Conference of Doble Clients, March 25 – 30, 2012, Boston, MA.

Campbell, R. J. (2011, October 28). "Regulatory incentives for electricity transmissions: Issues and cost concerns." Congressional Research Service. Retrieved April 1, 2014 from http://www.ieeeusa.org/policy/ eyeonwashington/2011/documents/electrans.pdf

Clark, Cammy. (2012, July 2). "Massive transformer makes long voyage to FPL's Turkey Point nuclear plant." *Miami Herald*. Retrieved April 1, 2014 from http://www.miamiherald.com/2012/06/06/2878497/massive-transformer-makes- long.html#storylink=cpy

Edison Electric Institute. (2005, May). "EEI Survey of Transmission Investment." Retrieved April 1, 2014 from http://www.eei.org/ourissues/ ElectricityTransmission/Documents/Trans_Survey_Web.pd f

Electromagnetic Pulse Commission. (2008, April). "Report of the Commission to Assess the Threat to the United States from Electromagnetic Pulse

(EMP) Attack." Retrieved January 3, 2014 from http://www.empcommission.org/docs/A2473-EMP_Commission-7MB.pdf

Federal Energy Regulatory Commission. (2014, January 16). "FERC Proposes Reliability Standard on Geomagnetic Disturbances." Retrieved April 3, 2014 from http://www.ferc.gov/media/news-releases/2014/2014-1/01-16-14-E-3.asp#.Uz1tKHPD8dV

Federal Energy Regulatory Commission. (2014, March 7). "Reliability Standards for Physical Security Measures." Order Directing Filing for Standards. Retrieved April 1, 2014 from http://www.ferc.gov/CalendarFiles/20140307185442-RD14-6-000.pdf

Global Data. (2009). "Power Transformers Market Analysis to 2020." Retrieved April 1, 2014 from http://www.articlesnatch.com/Article/Power-Transformers-Market-Analysis-To-2020-/1861724

Global Industry Analysts. (2011, October 12). "Global Electricity Transformers Market to Reach US$53.6 billion, 9.7 million Units by 2017." Retrieved April 1, 2014 from http://tdworld.com/business/global-transformer-market-1011/

Gouldon Reports. (2009, May 15). "World Transformer Markets 2002 to 2012." Retrieved April 1, 2014 from http://www.leonardo-energy.org/webfm_send/2731

Hyundai Heavy Industries. (2011, November 21). "Hyundai Heavy Completes Transformer Factory in Alabama." Retrieved April 1, 2014 from http://hyundaiheavy.com/news/view?idx=53

Indian Department of Heavy Industries. "Final report on the Indian capital goods industry." Retrieved April 1, 2014 from http://dhi.nic.in/indian_capital_goods_industry.pdf

International Energy Agency. (2013, November 12). World Energy Outlook. Retrieved April 1, 2014 from http://www.worldenergyoutlook.org/media/weowebsite/2013/LondonNovember12.pdf

Kappenman, J. (2010, January). "Geomagnetic storms and their impacts on the U.S. power grid." Retrieved January 3, 2014 from http://www.fas.org/irp/eprint/geomag.pdf

Metal Center News. (2011, May 4). "World steel demand to hit new record in 2012." Retrieved January 3, 2014 from http://www.metalcenternews.com/Editorial/SearchBackIssues/2011Issues/MCNMay2011/542011NewsWorldSteelDemandHitNewRecord/tabid/5078/Default.aspx

Mitsubishi Electric. (2013, April 22). "Mitsubishi Electric Transformer Factory Starts Operating in Memphis." Retrieved April 1, 2014 from http://www.mitsubishielectric.com/news/2013/0422.html

National Infrastructure Advisory Council. (2010, October 16). "A Framework for Establishing Critical Infrastructure Resilience Goals." Retrieved April 1, 2014 from http://www.dhs.gov/xlibrary/assets/niac/niac-a-framework-for-establishing-critical- infrastructure-resilience-goals-2010-10-19.pdf

North American Electric Reliability Corporation (NERC). (2010, June). "High-Impact, Low- Frequency Event Risk to the North American Bulk Power System." Retrieved April 1, 2014 from http://www.nerc.com/pa/CI/Resources/Documents/HILF_Report.pdf

NERC. Electricity Subsector Coordinating Council. (2010, November). "Critical Infrastructure Strategic Roadmap."

NERC. (2011, October). "Special report: Spare Equipment Database System." Retrieved April 1, 2014 from http://www.nerc.com/docs/pc/sedtf/SEDTF_Special_Report_October_2011.pdf

NERC. "Electricity Supply & Demand." Retrieved April 1, 2014 from http://www.nerc.com/pa/RAPA/ESD/Pages/default.aspx

NERC. "Project 2013-03 Geomagnetic Disturbance Mitigation." Retrieved April 1, 2014 from http://www.nerc.com/pa/Stand/Pages/Project-2013-03-Geomagnetic-Disturbance- Mitigation.aspx

Organization for Economic Cooperation and Development. "Members and Partners." Retrieved January 3, 2014 from http://www.oecd.org/about/membersandpartners/list-oecd-member- countries.htm

PR Newswire. (2011, July 26). "Frbiz Analyzes China's Power Transformer Market." Retrieved December 14, 2011 from http://www.prnewswire.com/news-releases/frbiz-analyzes-   chinas-power-transformer-market-99231019.html

PR Newswire. (2013, November 2). "Power Transformers Market - Global Industry Analysis, Size, Share, Growth, Trends and Forecast, 2013 – 2019," Transparency Market Research. Retrieved April 1, 2014 from http://www.prweb.com/releases/2013/11/prweb11294070.htm

Prolec GE. (2008, November). "DOE Distribution Transformer Efficiency Regulation Evaluation of Impact on the Industry." Retrieved April 1, 2014 from http://www.ieeerepc.org/documents/ProlecGEDOETransformerEfficiencyStandardsREP C.ppt

Reuters. (2011, September 18). "Research and Markets: China's Power Transformer Industry Report, 2010–2011." Retrieved April 1, 2014 from http://www.reuters.com/article/2011/09/19/idUS19990+19-Sep-2011+BW20110919

Schumacher, J. (2006, May 1). "Buying Transformers." *Transmission & Distribution World*. Retrieved April 1, 2014 from http://tdworld.com/business/power_buying_transformers/

Siemens AG, Power Transmission Division. (2011, September 19). "HVDC Growth Market – More Energy Highways for Europe's Power Grid." Retrieved April 1, 2014 from http://www.siemens.com/press/pool/de/events/2011/energy/2011-09- mallorca/presentation-niehage-e.pdf

Siemens. "Transformer Lifecycle Management." Retrieved April 1, 2014 from http://www.energy.siemens.com/mx/pool/hq/services/power-transmission-distribution/transformer-lifecycle-management/TLM_EN_.pdf

Siemens. (2010). "Siemens Transformers China." Retrieved April 1, 2014 from http://www.energy.siemens.com.cn/CN/powerTransmission/Transformers/Documents/T R%20China%20catalog_EN_2010.pdf

SPX. (2011, May 16). "Electrical Products Group." Retrieved April 1, 2014 from http://phx.corporate- ir.net/External.File?item=UGFyZW50SU Q9NDI3NDYzfENoaWxkSUQ9NDQ0NTg1fF R5cGU9MQ==&t=1

SPX. (2012, April 12). "SPX Unveils Newly Expanded SPX Transformer Solutions Manufacturing Facility in Waukesha, Wisconsin." Retrieved April 1, 2014 from http://www.spxtransformersolutions.com/large_power.html

SPX. (2013, September 11). "SPX Transformer Solutions Analyst Day Presentation."

State Grid Corporation of China. "Corporate profile." Retrieved January 5, 2014 from http://www.sgcc.com.cn/ywlm/aboutus/profile.shtml

The White House. (2010, May). "National Security Strategy." Retrieved April 1, 2014 from http://www.whitehouse.gov/sites/default/ files/rss_viewer/national_security_strategy.pdf

The White House. (2012, January 25). "National Strategy for Global Supply Chain Security." Retrieved February 5, 2014 from http://www. whitehouse.gov/sites/default/files/national_strategy_for_global_supply_ch ain_security.pdf

The White House (2013, February 12). "The Presidential Policy Directive— Critical Infrastructure Security and Resilience (PPD-21)." Retrieved April 1, 2014 from http://www.whitehouse.gov/the-press-office/2013/02/12/presidential-policy-directive- critical-infrastructure-security-and-resil

U.S. Congress, Office of Technology Assessment, Physical Vulnerability of Electric System to Natural Disasters and Sabotage, OTA-E-453 (Washington, DC: U.S. Government Printing Office, June 1990).

U.S. Department of Energy (DOE). (2006, August). "Benefits of using mobile transformers and mobile substations for rapidly restoring electrical service: A report to the United States Congress pursuant to Section 1816 of the Energy Policy Act of 2005." Retrieved April 1,2014 from http://energy/DocumentsandMedia/MTS_Report_to_Congress_FINAL_73 106.pdf

DOE, Office of Energy Efficiency and Renewable Energy. (2007, September). "Distribution transformers final rule technical support document (app. 3A). Core steel market analysis." Retrieved April 1, 2014 from https://www1.eere.energy.gov/buildings/appliance_standards/commercial/pdfs/tr ansform er_preanalysis_app3a.pdf

U.S. Department of Homeland Security (DHS). (2003, December). Homeland Security Presidential Directive-7 (HSPD-7). Retrieved January 4, 2012 from http://www.dhs.gov/xabout/laws/gc_1214597989952.shtm

DHS. (2013). "2013 NIPP: Partnering for Critical Infrastructure Security and Resilience," Retrieved April 1, 2014 from http://www.dhs.gov/sites/ default/files/publications/NIPP%202013_Partnering%20for%20   Critical %20Infrastructure%20Security%20and%20Resilience_508_0.pdf

DHS and DOE. (2010). "Energy Sector-Specific Plan." Retrieved April 1, 2014 from http://energy.gov/sites/prod/files/oeprod/DocumentsandMedia/ Energy_SSP_2010.pdf

DHS. "Power Hungry: Prototyping Replacement EHV Transformers." Retrieved April 1, 2014 from http://www.dhs.gov/files/programs/st-snapshots-prototyping-replacement-ehv- transformers.shtm

U.S. Energy Information Administration. (2011, September). "2011 International Energy Outlook." Retrieved January 4, 2014 from http://www.eia.gov/forecasts/ieo/ieo_tables.cfm

U.S. International Trade Commission (USITC). (2011, September). "Large power transformers from Korea, Publication 4256." Retrieved December 11, 2012 from http://www.usitc.gov/publications/701_731/Pub4256.pdf

USITC. (2011, August 4). "Conference Hearing for Investigation No.731-TA-1189 in the Matter of Large Power Transformers from Korea." Retrieved April 1, 2014 from http://www.usitc.gov/trade_remedy/731_ad_701_cvd/ investigations/2011/large_power_transformers/preliminary/PDF/Conferen ce%2008-04-2011.pdf

USITC. (2013, November). "Grain-Oriented Electrical Steel from China, Czech Republic, Germany, Japan, Korea, Poland, and Russia," Publication 4439. Retrieved April 1, 2014 from http://www.usitc.gov/publications/ 701_731/pub4439.pdf

USITC. "Interactive Tariff and Trade Dataweb." Retrieved April 1, 2014 from http://dataweb.usitc.gov/scripts/user_set.asp

Wald, M. (2012, March 14). "A Drill to Replace Crucial Transformers (Not the Hollywood Kind)." New York Times. Retrieved April 1, 2014 from http://www.nytimes.com/2012/03/15/business/energy-environment/ electric-industry- runs-transformer-replacement-test.html?_r=3

Working Group for Investment in Reliable and Economic Electric Systems and the Brattle Group. (2011, May). "Employment and economic benefits of transmission infrastructure investment in the U.S. and Canada." Retrieved December 1, 2013 from http://www.wiresgroup.com/images/ Brattle-WIRES_Jobs_Study_May2011.pdf

End Notes

[1] Throughout this report, the term large power transformer (LPT) is broadly used to describe a power transformer with a maximum capacity rating greater than or equal to 100 MVA unless otherwise noted.

[2] Throughout this report, the term large power transformer (LPT) is broadly used to describe a power transformer with a maximum capacity rating of 100 MVA or higher, unless otherwise noted. See Section 1.3, Scope and Definition of Large Power Transformers, for discussions related to the inconsistencies in the industry's definition of LPTs, and see Section 2.2, Physical Characteristics of Large Power Transformers, for key physical attributes of LPTs.

[3] In this report, the Electricity Sector refers to the electricity industry as described in the" 2010 Energy Sector- Specific Plan" (SSP). The Energy Sector, as delineated by Homeland Security Presidential Directive 7 (HSPD-7), includes the production, refining, storage, and distribution of oil, gas, and electric power, except for hydroelectric and commercial nuclear power facilities. The Energy Sector is not monolithic; it contains many interrelated industries that support the exploration, production, transportation, and delivery of fuels and electricity to the U.S. economy. See the 2010 Energy SSP, http://energy.gov/sites/prod/ files/oeprod/DocumentsandMedia/Energy_SSP_2010.pdf (accessed April 7, 2014).

[4] The Presidential Policy Directive—Critical Infrastructure Security and Resilience (PPD-21) aims to advance a national unity of efforts to strengthen and maintain secure, functioning, and resilient critical infrastructure. It also identified the Nation's 16 critical infrastructure sectors, and the agencies that would be responsible for carrying out the policy for each sector. DOE is the Sector-Specific Agency that is responsible for the coordination of critical infrastructure protection activities for the Energy Sector. See the full text of PPD-21 at http://www.whitehouse.gov/the-press-office/2013/02/12/presidential-policy-directive- critical-infrastructure-security- and-resil (accessed January 4, 2014).

[5] Such concern was raised in a June 1990 Congressional Report. See U.S. Congress, Office of Technology Assessment, Physical Vulnerability of Electric System to Natural Disasters and Sabotage, OTA-E-453 (Washington, DC: U.S. Government Printing Office, June 1990).

[6] "The National Security Strategy," The White House, May 2010, http://www.whitehouse.gov/ sites/default/files/rss_viewer/national_security_strategy.pdf (accessed December 15, 2013).

[7] See the "2013 NIPP: Partnering for Critical Infrastructure Security and Resilience," U.S. Department of Homeland Security (DHS), http://www.dhs.gov/sites/default/files/

publications/NIPP%202013_Partnering%20for%20Critical%20Infrastructure%20Security%20and%20Resilience_508_0.pdf; "National Strategy for Global Supply Chain Security," The White House, January 25, 2012, http://www.whitehouse.gov/sites/default/files/national_strategy_for_global_supply_chain_security.pdf; 2010 Energy SSP http://energy.gov/ ites/prod/files/oeprod/DocumentsandMedia/Energy_SSP_2010.pdf (all accessed March 1, 2014).

[8] "A Framework for Establishing Critical Infrastructure Resilience Goals," the National Infrastructure Advisory Council, October 16, 2010, http://www.dhs.gov/xlibrary/assets/niac/niac-a-framework-for-establishing-critical-infrastructure-resilience-goals-2010-10-19.pdf (accessed December 8, 2013).

[9] Ibid.

[10] "High-Impact, Low-Frequency Event Risk to the North American Bulk Power System," NERC, June 2010, http://www.nerc.com/pa/CI/Resources/Documents/HILF_Report.pdf (accessed March 26, 2014).

[11] "Critical Infrastructure Strategic Roadmap," Electricity Sub-sector Coordinating Council, November 2010.

[12] The Electricity Subsector Coordinating Council represents the Electricity Sector as described in the Energy Sector Specific Plan, which includes bulk power system entities defined by Section 215 of the Federal Power Act.

[13] "Benefits of Using Mobile Transformers and Mobile Substations for Rapidly Restoring Electrical Service," a report to the United States Congress pursuant to Section 1816 of the Energy Policy Act of 2005, U.S. Department of Energy, August 2006, http://energy.gov/sites/prod/files/oeprod/DocumentsandMedia/MTS_Report_to_Congress_FINAL_73106.pdf (accessed March 26, 2014).

[14] Ibid. This study does not consider distribution transformers in the assessment, because the United States maintains domestic manufacturing capacity and backup supplies of distribution transformers. Moreover, a localized power outage at the distribution level does not present significant reliability threats, and utilities often maintain spare transformer equipment of this size range.

[15] Ibid.

[16] "Large Power Transformers from Korea," U.S. International Trade Commission (USITC), Publication 4256, September 2011, http://www.usitc.gov/publications/701_731/ Pub4256.pdf (accessed December 11, 2013).

[17] "Special Report: Spare Equipment Database System," the North American Electric Reliability Corporation, October 2011, http://www.nerc.com/docs/pc/sedtf/SEDTF_Special_Report_October_2011.pdf (accessed November 22, 2013).

[18] 2013 NERC Electricity Supply & Demand Database, http://www.nerc.com/pa/RAPA/ESD Pages/default.aspx (accessed March 26, 2014).

[19] Electricity is generally produced at 5 to 34.5 kV and distributed at 15 to 34.5 kV, but transmitted at 115 to 765 kV for economical, low-loss, long-distance transmission on the grid.

[20] "Large Power Transformers from Korea," USITC, Publication 4256, September 2011.

[21] "Large Power Transformers from Korea," USITC, Publication 4256, September 2011, p. I-5.

[22] "Benefits of Using Mobile Transformers and Mobile Substations for Rapidly Restoring Electrical Service," U.S. Department of Energy, 2006.

[23] "Special Report: Spare Equipment Database System," NERC, October 2011.

[24] Conference Hearing for Investigation No.731-TA-1189, In the Matter of Large Power Transformers from Korea, USITC, August 4, 2011, http://www.usitc.gov/trade_remedy/731_ad_701_cvd/investigations/2011/large_power_transformers/preliminary/PDF/Conference%2008-04-2011.pdf (accessed March 26, 2014).

[25] "Large Power Transformers from Korea," USITC, Publication 4256, September 2011.

[26] Ibid., p. V-I.

[27] Conference Hearing for Investigation No.731-TA-1189, USITC, August 4, 2011, pp. 135–136.

[28] "Large Power Transformers from Korea," USITC, Publication 4256, September 2011, pp. I-9–I-10.

[29] Conference Hearing for Investigation No.731-TA-1189, USITC, August 4, 2011, p. 95.

[30] Ibid., p. II-7.

[31] SPX Transformer Solutions Analyst Day Presentation, September 11, 2012.

[32] Industry source estimate.

[33] "Large Power Transformers from Korea," USITC, Publication 4256, September 2011, p. II-7.

[34] Siemens, Transformer Lifecycle Management, http://www.energy.siemens.com/mx/pool/hq/services/power-transmission-distribution/transformer-lifecycle-management/TLM_EN_.pdf (accessed March 26, 2014).

[35] Conference Hearing for Investigation No.731-TA-1189, USITC, August 4, 2011, p. 100.

[36] "Large Power Transformer and Schnabel Rail Car," Business and Technical Report, National Security Industrial Readiness, Pre-decisional Draft for Official Use Only.

[37] "Railcar rental program for power transformer relocation," ABB, http://www05.abb.com/global/scot/scot252.nsf/veritydisplay/36bcc4e173d5c1558525760b00711641/$file/1zul0046 05-300_railcar_r4.pdf (accessed March 26, 2014).

[38] Clark, Cammy, "Massive transformer makes long voyage to FPL's Turkey Point nuclear plant," July 2, 2012, Miami Herald, http://www.miamiherald.com/2012/06/06/2878497/massive-transformer-makes- long.html#storylink=cpy (accessed April 1, 2014).

[39] "Large Power Transformers from Korea," USITC, Publication 4256, September 2011, p. V-1, and an industry source estimate.

[40] Ibid.

[41] Ibid., p. VI-1.

[42] "Grain-Oriented Electrical Steel from China, Czech Republic, Germany, Japan, Korea, Poland, and Russia," USITC, Publication 4439, November 2013, http://www.usitc.gov/publications/701_731/pub4439.pdf (accessed March 25, 2014).

[43] "Flux density" generally refers to the total number of magnetic lines of force per unit area (i.e., the density of magnetic lines of force, or magnetic flux lines, passing through a specific area.) Source: Ibid.

[44] "Grain-Oriented Electrical Steel from China, Czech Republic, Germany, Japan, Korea, Poland, and Russia," USITC, Publication 4439, November 2013.

[45] Ibid.

[46] Conference Hearing for Investigation No.731-TA-1189, USITC, August 4, 2011, p. 97.

[47] Estimates provided by an industry source.

[48] "Appendix 3A. Core Steel Market Analysis, Technical Support Document: Energy Efficiency Program for Consumer Products and Commercial and Industrial Equipment Distribution Transformers," Office of Energy Efficiency & Renewable Energy (EERE), DOE, April 2013, https://www1.eere.energy.gov/buildings/appliance_standards/commercial/pdfs/transformer_preanalysis_app3a.pdf (accessed December 30, 2013).

[49] "Technical Support Document: Energy Efficiency Program for Consumer Products and Commercial and Industrial Equipment Distribution Transformers," EERE, DOE, April 2013.

[50] "World Steel in Figures 2013," Worldsteel Association, http://www.worldsteel.org/dms/internetDocumentList/bookshop/Word-Steel-in-Figures-2013/document/World%20Steel%20in%20Figures%202013.pdf (accessed April 9, 2014).

[51] Technical Support Document: Energy Efficiency Program for Consumer Products and Commercial and Industrial Equipment Distribution Transformers," EERE, DOE, April 2013.

[52] Schumacher, Judson, "Buying Transformers," Transmission & Distribution World, May 1, 2006,http://tdworld.com/business/power_buying_transformers/ (accessed March 26, 2014).

[53] "Large Power Transformers from Korea," USITC, Publication 4256, September 29, 2011.

[54] "2013 International Energy Outlook," U.S. Energy Information Administration (EIA), http://www.eia.gov/forecasts/ieo/ieo_tables.cfm (accessed January 8, 2014).

[55] The EIA's 2011 IEO data are divided according to Organization for Economic Cooperation and Development members (OECD) and nonmembers (non-OECD). The 34 OECD member countries as of March 2014 are listed here: http://www.oecd.org/about/membersandpartners/list-oecd-member-countries.htm (accessed January 3, 2014).

[56] "2013 International Energy Outlook," EIA.

[57] "2013 World Energy Outlook," International Energy Agency, November 12, 2013, http://www.worldenergyoutlook.org/media/weowebsite/2013/LondonNovember12.pdf (accessed January 8, 2014).

[58] The EEI is an association of U.S. shareholder-owned electric power companies. Its members serve 95 percent of the ultimate customers in the shareholder-owned segment of the industry and represent approximately 70 percent of the U.S. electric power industry. See http://www.eei.org/Pages/default.aspx (accessed December 1, 2013).

[59] Campbell, Richard J., "Regulatory Incentives for Electricity Transmissions – Issues and Cost Concerns," Congressional Research Service, October 28, 2011, http://www.ieeeusa.org/policy/eyeonwashington/2011/documents/electrans.pdf (accessed March 26, 2014).

[60] EEI Survey of Transmission Investment, EEI, May 2005, http://www.eei.org/ourissues/ElectricityTransmission/Documents/Trans_Survey_Web.pdf (accessed March 26,2014).

[61] "Employment and Economic Benefits of Transmission Infrastructure Investment in the U.S. and Canada," WIRES (Working Group for Investment in Reliable and Economic Electric Systems) In Conjunction with The Brattle Group, May 2011, http://www.wiresgroup.com/images/Brattle-WIRES_Jobs_Study_May2011.pdf (accessed December 1, 2013).

[62] Ibid.

[63] EEI Survey of Transmission Investment, EEI, May 2005.

[64] "World Transformer Markets 2002 to 2012," Gouldon Reports 2009, Presented to Leonardo Energy, M ay 15,2009, http://www.leonardo-energy.org/sites/leonardo-energy/files/root/pdf/2009/20090515-trafomarket.pdf (accessed March 26, 2014).

[65] Ibid.

[66] "Power Transformers Market Analysis to 2020," Global Data, 2009, http://www.articlesnatch.com/Article/Power-Transformers-Market-Analysis-To-2020-/1861724 (accessed March 26, 2014).

[67] "Power Transformers Market - Global Industry Analysis, Size, Share, Growth, Trends and Forecast, 2013 –2019," Transparency Market Research, November 2, 2013, http://www.prweb.com/releases/2013/11/prweb11294070.htm (accessed December 12, 2013).

[68] "Power Transformers Market Analysis to 2020," Global Data, 2009.

[69] "Global Electricity Transformers Market To Reach US$53.6 Billion, 9.7 Million Units by 2017," Global Industry Analysts, October 12, 2011, http://tdworld.com/business/global-transformer-market-1011/ (accessed March 26, 2014).

[70] Global Industry Analysts, October 12, 2011.

[71] Global Industry Analysts, Inc., October 5, 2011; "Shareholders Meeting," Alstom, June 28, 2011; ABB Strategy 2011, Sept. 5, 2007.

[72] ABB: http://www.eyeproject.cl/files/94151d_ABB-BR-31-TrafoStar-FlipChart12Reasons.pdf; Siemens: http://www.energy.siemens.com.cn/CN/powerTransmission/Transformers/Documents/TR%20China%20catalog_E N_2010.pdf; Alstom Grid: http://www.alstom.com/grid/news-and-events/press-releases/upgrade-project-at-Alstom- Grid-s-Rocklea-factory/ (all accessed December 3, 2013).

[73] Alstom Grid: http://www.alstom.com/grid/solutions/high-voltage-power-products/electrical-power-transformers/ (accessed December 3, 2013); Siemens does not disclose its total annual production capacity; however, it has an annual manufacturing capacity of 70,000 MVA in China.

[74] "Final Report on the Indian Capital Goods Industry," http://dhi.nic.in/indian_capital_goods_industry.pdf (accessed December 14, 2013).

[75] "Benefits of Using Mobile Transformers and Mobile Substations for Rapidly Restoring Electrical Service," DOE, August 2006. This DOE source provides an estimated voltage range of an LPT as 115–765 kV and the estimated power range as 200–1,200 MVA. "Report of the Commission To Assess the Threat to the United States from Electromagnetic Pulse (EMP) Attack," EMP Commission, April 2008, http://www.empcommission.org/docs/A2473-EMP_Commission-7MB.pdf (accessed January 3, 2014). Kappenman, John, "Geomagnetic Storms and Their Impacts on the U.S. Power Grid." Prepared for Oak Ridge National Laboratory, January 2010, http://www.fas.org/irp/eprint/geomag.pdf (accessed January 3, 2014).

[76] "Large Power Transformers from Korea," USITC, Publication 4256, September 2011.

[77] Figure 14 was derived from the USITC's Publication 4256 on "Large Power Transformers from Korea," which defined an LPT as a power transformer with a capacity rating greater than or equal to 60 MVA. China's LPT data was derived from an industry analysis as reported in Reuters, which described used transformers with a voltage rating of 220 kV and above to describe the market. More recent data is not available.

[78] In this paper, the term annual production capacity represents the capability of a manufacturer to produce transformers of all types and sizes, ranging from small to large, in an annual basis.

[79] Conference Hearing for Investigation No.731-TA-1189, USITC, August 4, 2011, p. 113.

[80] "Large Power Transformers from Korea," USITC, Publication 4256, September 2011.

[81] Figure 15 represents LPT manufacturing plants in North America as of May 2012. Plants became operational after 2010 were not included as part of the 2010 manufacturing capacity analysis in section 5.2.

[82] Efacec, http://www.efacecusa.com/; Terracon Company, http://www.terracon.com/projects/efacec-transformers/ (accessed April 3, 2014).

[83] "Hyundai Heavy Completes Transformer Factory in Alabama," Hyundai Heavy Industries, November 21, 2011, http://hyundaiheavy.com/news/view?idx=53 (accessed March 26, 2014).

[84] "Mitsubishi Electric To Build Transformer Factory in Memphis," Mitsubishi Press Release, February 14, 2011, http://www.mitsubishielectric.com/news/2011/0214-a.html (accessed December 2, 2014).

[85] "SPX Unveils Newly Expanded SPX Transformer Solutions Manufacturing Facility in Waukesha, Wisconsin," April 12, 2012, http://www.spxtransformersolutions.com/large_power.html (accessed March 26, 2014).

[86] Ibid.; "Large Power Transformers," SPX, http://www.spxtransformersolutions.com/assets/documents/LP%20Brochure_web.pdf (accessed December 22, 2013).

[87] See "Appendix C. Historical Imports of Large Power Transformers in the United States" for data used in Figure 16.

[88] USITC Interactive Tariff and Trade DataWeb, http://dataweb.usitc.gov/scripts/user_set.asp (accessed March 26, 2014).

[89] Ibid.; HHI Electro Electric Systems.

[90] SPX Electrical Products Group, May 2011 and an industry source estimate.

[91] "September 2012 Investor Presentation," SPX Transformer Solutions, September 2012.

[92] Schumacher, Judson, "Buying Transformers," Transmission & Distribution World, May 1, 2006, http://tdworld.com/business/power_buying_transformers/index.html (accessed January 11, 2014).

[93] Bartley, William H., "Analysis of Transformer Failures," Hartfort Steam Boiler Inspection & Insurance Co., 79th International Conference of Doble Clients, March 25 – 30, 2012, Boston, MA.

[94] Ibid.

[95] Conference Hearing for Investigation No.731-TA-1189, USITC, August 4, 2011, pp. 147–148.

[96] Bartley, W.H., Hartford Steam Boiler Inspection & Insurance Co., 2012.

[97] Ibid.

98 "Reliability Standards for Physical Security Measures," FERC, Order Directing Filing for Standards, March 7, 2014, http://www.ferc.gov/CalendarFiles/20140307185442-RD14-6-000.pdf (accessed April 1, 2014).

99 "FERC Proposes Reliability Standard on Geomagnetic Disturbances," FERC, January 16, 2014, http://www.ferc.gov/media/news-releases/2014/2014-1/01-16-14-E-3.asp#.Uz1tKH PD8dV (accessed April 3, 2014).

100 "Project 2013-03 Geomagnetic Disturbance Mitigation," NERC, http://www.nerc.com/pa/ Stand/Pages/Project-2013-03-Geomagnetic-Disturbance-Mitigation.aspx (accessed March 24, 2014).

101 For more information, see http://www.eei.org/issuesandpolicy/transmission/Pages/ sparetransformers.aspx and http://www.nerc.com/pa/RAPA/sed/Pages/Spare-Equipment-Database-(SED).aspx (accessed March 25, 2014).

102 For more information about RecX, see: http://www.dhs.gov/files/programs/st-snapshots-prototyping-replacement-ehv-transformers.shtm and http://www.nytimes.com/2012/03/ 15/business/energy-environment/electric- industry-runs-transformer-replacement-test.html? _r=1 (both accessed March 24, 2013).

103 Corporate Profile, State Grid of Corporation of China, http://www.sgcc.com.cn/ywlm/ aboutus/profile.shtml (accessed January 5, 2014). Another state-owned enterprise, China Southern Power Grid manages the power transmission and distribution systems in the remaining five provinces on the mainland.

104 "Research and Markets: China's Power Transformer Industry Report, 2010–2011," Reuters, September 18, 2011, http://www.reuters.com/article/2011/09/19/idUS19990+19-Sep-2011+ BW20110919 (accessed December 14, 2013).

105 Ibid.

106 Ibid.

107 "Frbiz Analyzes China's Power Transformer Market," PR News, July 26, 2011, http://www.prnewswire.com/news-releases/frbiz-analyzes-chinas-power-transformer-market-99231019.html (accessed December 14, 2013).

108 "Siemens Transformers China," 2010, http://www.energy.siemens.com.cn/CN/power Transmission/Transformers/Documents/TR%20China%20catalog_E N_2010.pdf (accessed December 21, 2013).

In: Large Power Transformers
Editor: Vincent Beck

ISBN: 978-1-63463-270-6
© 2015 Nova Science Publishers, Inc.

Chapter 2

PHYSICAL SECURITY OF THE U.S. POWER GRID: HIGH-VOLTAGE TRANSFORMER SUBSTATIONS[*]

Paul W. Parfomak

SUMMARY

In the United States, the electric power grid consists of over 200,000 miles of high-voltage transmission lines interspersed with hundreds of large electric power transformers. High voltage (HV) transformer units make up less than 3% of transformers in U.S. power substations, but they carry 60%-70% of the nation's electricity. Because they serve as vital nodes and carry bulk volumes of electricity, HV transformers are critical elements of the nation's electric power grid. HV transformers are also the most vulnerable to intentional damage from malicious acts. Recent security exercises, together with a 2013 physical attack on transformers in Metcalf, CA, have focused congressional interest on the physical security of HV transformers. They have also prompted new grid security initiatives by utilities and federal regulators. Legislative proposals, notably the Grid Reliability and Infrastructure Defense Act (H.R. 4298 and S. 2158), would expand these efforts by strengthening federal authority to secure the U.S. grid.

---

[*] This is an edited, reformatted and augmented version of a Congressional Research Service publication, No. R43604, dated June 17, 2014.

For more than 10 years, the electric utility industry and government agencies have engaged in a number of initiatives to secure HV transformers from physical attack and to improve recovery in the event of a successful attack. These initiatives include coordination and information sharing, spare equipment programs, security standards, grid security exercises, and other measures. There has been some level of physical security investment and an increasing refinement of voluntary grid security practices across the electric power sector for at least the last 15 years. Several major transmission owners have recently announced significant new initiatives specifically to improve the physical security of critical transformer substations in light of the Metcalf attack.

On March 7, 2014, the Federal Energy Regulatory Commission (FERC) ordered the North American Electric Reliability Corporation (NERC) to submit to the Commission new reliability standards requiring certain transmission owners "to take steps or demonstrate that they have taken steps to address physical security risks and vulnerabilities related to the reliable operation" of the power grid. In its order, FERC states that physical security standards are necessary because "the current Reliability Standards do not specifically require entities to take steps to reasonably protect against physical security attacks." According to FERC's order, the new reliability standards will require grid owners to perform risk assessments to identify their critical facilities, evaluate potential threats and vulnerabilities, and implement security plans to protect against attacks.

There is widespread agreement among state and federal government officials, utilities, and manufacturers that HV transformers in the United States are vulnerable to terrorist attack, and that such an attack potentially could have catastrophic consequences. But the most serious, multi-transformer attacks would require acquiring operational information and a certain level of sophistication on the part of potential attackers. Consequently, despite the technical arguments, without more specific information about potential targets and attacker capabilities, the true vulnerability of the grid to a multi-HV transformer attack remains an open question. Incomplete or ambiguous threat information may lead to inconsistency in physical security among HV transformer owners, inefficient spending of limited security resources at facilities that may not really be under threat, or deployment of security measures against the wrong threat.

As the electric power industry and federal agencies continue their efforts to improve the physical security of critical HV transformer substations, Congress may consider several key issues as part of its oversight of the sector: identifying critical transformers, confidentiality of critical transformer information, adequacy of HV transformer protection, quality of federal threat information, and recovery from HV transformer attacks.

INTRODUCTION[1]

The electric utility industry operates as an integrated system of generation, transmission, and distribution facilities to deliver electric power to consumers. In the United States, this system consists of over 9,000 electric generating units connected to over 200,000 miles of high-voltage transmission lines strung between large towers and rated at 230 kilovolts (kV)[2] or greater.[3] This network is interspersed with hundreds of large electric power transformers whose function is to adjust electric voltage as needed to move power across the network (Figure 1). High voltage (HV) transformer units make up less than 3% of transformers in U.S. power substations, but they carry 60%-70% of the nation's electricity.[4] Because they serve as vital transmission network nodes and carry bulk volumes of electricity, HV transformers are critical elements of the nation's electric power grid.

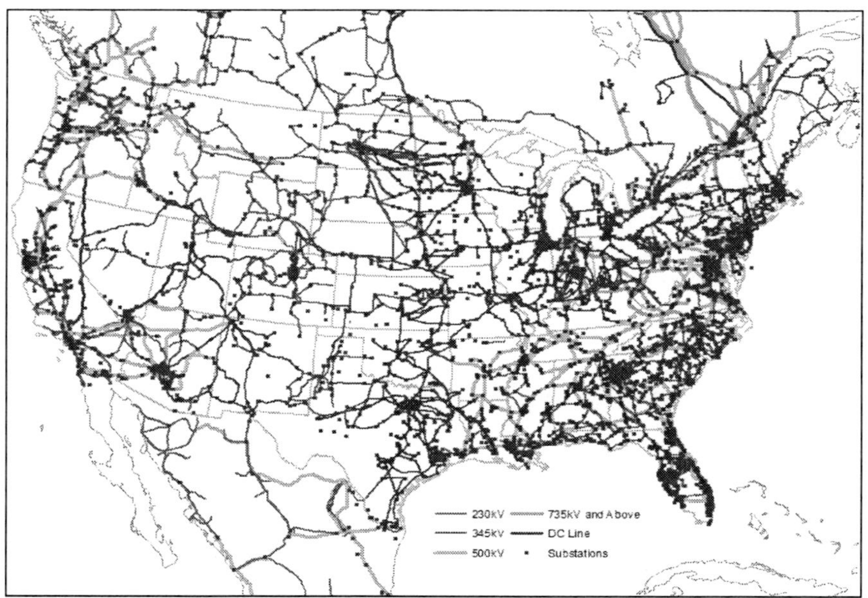

Sources: CRS analysis of GIS data from Platts, HSIP Gold 2013 (Ventyx), and Esri.

Figure 1. Electric Transmission Network.

The U.S. electric power grid has historically operated with such high reliability that any major disruption, either caused by weather, operational errors, or sabotage, makes news headlines. Such outages can have considerable negative impacts on business, government services, and daily life. Notwithstanding its high reliability overall, the U.S. power grid has periodically experienced major regional outages. Recent examples include the Northeast Blackout of 2003 (which affected 55 million customer in eight states and Canada) and extended outages in the New York/New Jersey area after Superstorm Sandy in 2012.

Congressional Interest

The various parts of the electric power system are all vulnerable to failure due to natural or manmade events. However, for reasons discussed below, HV transformers are considered by many experts to be the most vulnerable to intentional damage from malicious acts. Congress has long been concerned about grid security in general, but recent security exercises, together with a 2013 physical attack on transformers in Metcalf, CA, have focused congressional interest on the physical security of HV transformers, among other specific aspects of the grid.[5] They have also prompted new grid security initiatives by utilities and federal regulators. Recent legislative proposals, notably the Grid Reliability and Infrastructure Defense Act (H.R. 4298 and S. 2158), would expand these efforts by strengthening federal authority to secure the U.S. grid. The physical security of HV transformers and associated policy issues are the subject of this report.

HV TRANSFORMER RISKS AND VULNERABILITY

The main risk from a physical attack against the electric power grid—primarily towers and transformers—is a widespread power outage lasting for days or longer. Utilities regularly experience damage to transmission towers due to both weather and malicious activities and are able to recover from this damage fairly rapidly. Thus, while occasionally causing blackouts, physical attacks on towers generally have not resulted in widespread or long-lasting outages. Likewise, the power industry has experienced mechanical failure of individual HV transformers within a single control area resulting in blackouts lasting hours. However, no region in the United States has experienced simultaneous failures of multiple HV transformers. Experts have long asserted that a coordinated and simultaneous attack on multiple HV transformers could have severe implications for reliable electric service over a large geographic area, crippling its electricity network and causing widespread, extended blackouts. Such an event would have serious economic and social consequences. This section discusses in more detail HV transformer characteristics and physical security risks associated with them.

High Voltage Power Transformers

Utility transformers control the voltage of electricity so that it can be synchronized with other power supplies, transmitted long distances, and distributed to customers. Transformers range in size from small, pole-mounted units that may serve a dozen homes to transmission units that serve an entire city. The larger the transformer, the higher the voltage the transformer can handle.

Utility transformers, regardless of size, fundamentally consist of copper wire wrapped around a metallic "core" within an insulated protective housing covered with a 5/8 to 3/4-inch mild steel tank. They are linked to the power grid by protruding metal and (usually) ceramic connectors called "bushings" which resemble giant spark plugs. Larger transformers generate waste heat during operation, so they are cooled by a system of internally circulating oil and external radiators, analogous to the cooling system in a car engine. Transmission transformers are located in network substations along with transmission lines, associated electric equipment, and system controls. These substations may be found in remote locations or near urban centers, depending upon regional transmission needs. Many are located alongside electric generation plants, linking those plants to the grid.

Voltage Management in the U.S. Power System

Electricity produced at U.S. generating stations is converted into a set of three alternating electric currents called three-phase power.[6] The first step in delivering this power is transforming it from the generated voltage (typically 15- 50 kV) to higher voltage (138-765 kV), allowing transmission over long distances in greater volumes most efficiently (Figure 2).[7] This initial voltage step-up occurs by means of transformers located at transmission substations adjacent to the generating facilities. (The three phases of power are carried separately over three wires on transmission towers.)

Close to the ultimate consumer, the power is stepped-down at another transformer substation to lower voltages, typically 13 kV or less. At this point, the power is considered to have left transmission and entered the local distribution system.

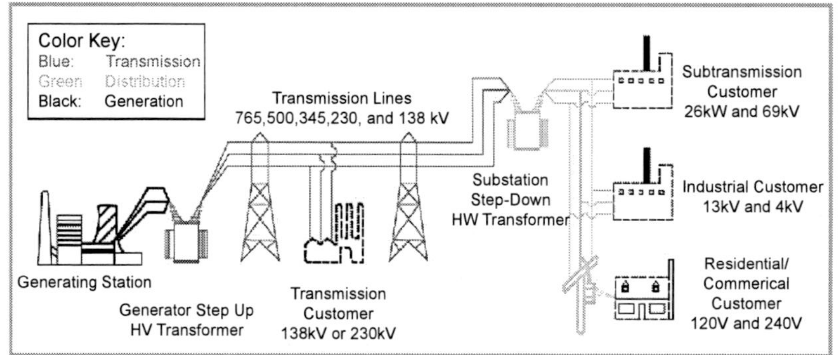

Source: Adapted by CRS from: U.S.-Canada Power System Outage Task Force, Final Report on the August 14, 2003 Blackout in the United States and Canada: Causes and Recommendations, April 2004, Figure 2.1.

Figure 2. Step-Up and Step-Down HV Transformers in the Grid.

Source: Courtesy of Pauwels Canada, Inc., 2003.

Figure 3. 345 kV Transformer Installation.

High-voltage transformers, especially units above 345 kV, are physically large and extraordinarily heavy. For example, Figure 3 shows a new 345 kV transformer many times larger than the pickup truck parked alongside. This transformer unit weighs 435 tons, including 29,000 gallons of cooling oil.[8] (Note that the vertical bushings are not yet connected to transmission lines because the unit is being moved.) This is a three-phase unit, with one bushing for each of the three phases. Some substations alternatively employ separate single-phase transformers in sets of three.

Generally, the higher the transformer's voltage, the larger the transformer. A three-phase 765kV transformer could be 45 feet tall and occupy a footprint of 2,200 square feet—about the size of an average new single-family house.[9]

Manufacture and Cost

Most HV transformers are unique and therefore are designed and manufactured to custom specifications for a specific network application. In 2010, the lead time between an HV transformer order and delivery ranged from 5 to 12 months for U.S. manufacturers and 6 to 16 months for foreign manufacturers, although lead times well over 20 months could be required in certain situations.[10] This process may include three to four months for the engineering design alone.[11] Since manufacturing generally occurs on a single production line with just-in-time component supplies, advanced production scheduling is important for managing delivery. Physical assembly is labor intensive, requiring manual winding of the copper wire around the transformer core and frequent engineering checks during manufacturing. Extensive testing of completed units also contributes to HV transformer manufacturing time.

The installed cost for an HV transformer depends heavily on its configuration and specific design requirements. New HV transmission substations can cost well in excess of $10 million, including the cost of transformers and other station equipment. According to the U.S. Department of Energy (DOE), the factory prices for HV transformers typically range from $2 million for a 230 kV unit to $7.5 million for a 765 kV unit, before transportation and installation costs.[12]

U.S. Manufacturing Capability

From 1950 to 1970, utility construction of large generation plants and associated transmission networks fueled a robust U.S. manufacturing market for large transformers. During this period, the United States (and Canada) accounted for approximately 40% of global demand for such units.[13] After 1970, however, utility investment in transmission infrastructure began falling off due to perceived overcapacity, public resistance to transmission siting, and greater regulatory scrutiny of capital expenditures. Beginning in the late 1980s, uncertainty about industry restructuring and the introduction of competition made grid owners even less willing to invest in new transmission. This decline in U.S. transmission investment greatly reduced domestic demand for large transformers, especially HV transformers. By the late 1990s, the United States and Canada accounted for only 20% of global large transformer sales.[14] Demand in the United States has subsequently increased, however. For example, between 2005 and 2013, the total value of large transformers (including medium- and high-voltage units) imported to the United States more than doubled, from $284 million (363 units) to $676 million (496 units).[15]

At the same time, global demand for transformers continued to grow and more foreign manufacturers entered the market. According to U.S. industry representatives, many of these foreign manufacturers benefited from dramatically lower labor costs, so they could underbid U.S. transformer makers for the remaining U.S. demand. Some of these foreign manufacturers may have been protected by import barriers which effectively closed their home markets to U.S. transformer imports. Today, there is limited manufacturing capacity in the United States for HV transformers. Five U.S. facilities state that they can manufacture transformers rated 345 kV or above, although it is not clear how many units in this range they have actually produced. Canada and Mexico have five additional HV manufacturing plants.[16] While limited domestic HV transformer manufacturing may increase delivery time, utilities have not reported difficulty in obtaining needed equipment.

HV Transformer Sites in the United States

There are several thousand HV transformers operating in the United States. Approximately 2,100 are very large units rated 345 kV and above.[17] Investor-owned utilities own most of these, although public utilities such as the Power Marketing Administrations (i.e., Bonneville Power Administration and Western Area Power Administration), Tennessee Valley Authority, and the Los Angeles Department of Water and Power own many HV transformers as well.[18] HV transformer substations are distributed throughout the electric grid, as shown in Figure 1, with the greatest number in the eastern part of the country.

Criticality of HV Transformers

Because they carry so much electricity, the destruction of HV transformers can seriously reduce the transmission capacity of a regional electric power grid and lead to extended blackouts. The impact of such a failure would depend on the electricity flows in that part of the grid, congestion from major network bottlenecks, and the status of other key facilities such as power plants, transmission lines, and other substations. Power grid planners generally anticipate the possible loss of a single HV transformer substation and are prepared to reroute power flows as necessary to maintain regional electric service. But the simultaneous loss of multiple HV transformers, especially in a constrained transmission area, could exceed the capability of a regional network to reroute power through secondary lines.[19]

Numerous publicly available studies have analyzed the risks of a multiple HV transformer failure. For example, the Congressional Office of Technology Assessment (OTA) in a 1990 report on the physical vulnerability of the electric power system found that

> In most cases, the nearly simultaneous destruction of two or three transmission substations would cause a serious blackout of a region or utility, although of short duration where there is an approximate balance of load and supply.... The destruction of more than three transmission substations would cause long-term blackouts in many areas of the country.[20]

In such an emergency scenario, limited electric service could likely be restored in the short term by imposing "rolling" blackouts, rerouting transmission, and using portable transformers. Nonetheless, the loss of key HV substations would leave the regional network crippled and highly susceptible to further disturbance and cascading failure.[21] According to power industry experts, certain parts of the U.S. transmission network are particularly vulnerable to HV substation disruption. These areas may have severely constrained transmission paths relying on a small number of HV transformers in extremely critical network locations. According to press accounts, a FERC power flow analysis in 2013 identified 30 such critical HV transformer substations across the continental United States; disabling as few as nine of these substations during a time of peak electricity demand reportedly could cause a "coast-to-coast blackout."[22] Not all industry experts agree on the potential severity and duration of a blackout from a multi-transformer attack, however, although it is generally accepted that severe outages may be technically possible.[23]

Physical Vulnerability of HV Transformers

All HV transformers are designed to withstand operational risks such as lightning strikes, hurricanes, and network power fluctuations—but they are vulnerable to intentional physical attacks. Despite their great size and internal complexity, HV transformers can be readily disabled or destroyed. According to one manufacturer, "if someone were to intentionally try ... it is a surprisingly simple task and there are a large number of ways to conceivably damage a transformer beyond repair."[24] Transformer experts have asserted that a bad actor with basic knowledge of transformer design could inflict irreparable damage. Such attacks can cause massive electrical short circuits and oil fires that would destroy an HV transformer and damage surrounding infrastructure. One fire at a 345 kV substation in Texas, for example, destroyed the transformer and burned for five hours, causing "plumes of smoke that could be seen for miles."[25] In addition to direct attacks on the transformers themselves, HV substations can be further disabled by damaging associated transmission lines or control centers that may be located on site.

Because HV transformers are so big and are connected to the largest overhead transmission towers, they are easily identified along major transmission corridors. High voltage transformers are usually housed in substations that are enclosed with a chain-link fence. Guards are not often stationed at these facilities under normal operating circumstances. Consequently, HV transformers are ordinarily easier to access than other critical electric facilities such as generation plants and control centers. Utilities use closed-circuit surveillance and other methods to detect intrusion. However, access to the substation may be achieved by either cutting or scaling the chain-link fence. Once inside, a saboteur could cause damage by accessing the control room or physically damaging the HV transformer. Penetrating the 5/8 to 3/4-inch steel tank with any device could short-circuit the windings and irreparably destroy the transformer. Alternatively, a saboteur could attempt to open a valve and drain the insulating oil. Igniting the oil might cause the transformer to arc and eventually explode. With a clear line of sight, an attacker could also disable transformers from a distance using conventional rifles.

The vulnerability of individual transformer substations has been demonstrated by successful attacks in recent years. In the most serious case, a rifle attack occurred in April 2013 at PG&E's 500 kV substation in Metcalf, CA. In this attack, multiple individuals outside the substation reportedly shot at the HV transformer radiators with .30 caliber rounds, causing them to leak cooling oil, overheat, and become inoperative.[26] In October 2013, the U.S. Justice Department charged an individual with attacks on the transmission grid in Arkansas, including a deliberate fire at Entergy's 500 kV substation in Lonoke County. The fire consumed the substation control house but electrical service was not interrupted.[27] In 2005, at a Progress Energy substation in Florida, a rifle attack ruptured a transformer oil tank, ultimately causing an explosion and local blackout.[28] Other attacks on substation equipment have been reported with some regularity, although most have been attributed to vandals or careless hunters.

It is very difficult to restore a damaged HV transformer substation. As noted above, transmission experts assert that most HV transformers currently in service are custom designed and, therefore, cannot be generally interchanged. Furthermore, at $3-5 million per unit or more, maintaining large inventories of spare HV transformers solely as emergency replacements is prohibitively costly, so limited extras are on hand. The number of spares a utility maintains is increasingly sensitive information, but one regional transmission control area reported in 2007 that it maintained 29 spares for 188 transformers rated 500 kV on its system.[29] Programs for the sharing of spare HV transformers among multiple utilities are discussed later in this report.

Within the United States, transportation of HV transformers is difficult. Due to their size and weight, most HV transformers are transported on special railcars, each with up to 36 axles to distribute the load. There are fewer than 20 of these railcars in the Unites States rated to carry 500 tons or more, which can present a logistical problem if they are needed in a transformer emergency.[30] Some specialized flatbed trucks can also carry heavy transformer loads over public roadways, but the few such trucks that exist have less carrying capacity and greater route restrictions than the railcars because HV transformers may exceed highway weight limits.

Targeting of HV Transformers

Malicious individuals could, without significant training, identify critical HV transformer locations and time an attack for greatest effect. This could be accomplished with basic knowledge of transmission operations and regional network characteristics drawn from publicly available sources, including electric marketing data indicating constrained areas of the network.[31] As stated in a 2012 National Research Council report, "terrorists could selectively target key equipment, especially large transformers."[32] The OTA report describes such a scenario:

> [One] example is a city served by eight transmission substations spread along a 250-mile line and located in five States. A knowledgeable saboteur would be needed to identify and find the eight transmission substations. A highly organized attack would also be required. However the damage would be enormous, blacking out a four-State region, with severe degradation of both reliability and economy for months.[33]

In 1997, the Irish Republican Army reportedly planned this kind of coordinated attack against six transmission substations in the United Kingdom. Although the attack was prevented, had it been successful it reportedly could have caused widespread power outages in London and the South East of England for months.[34]

It is relatively easy to learn about HV transformer vulnerabilities from engineers and operators experienced with this technology, either domestically or abroad, since the same technology is used in power grids throughout the world. In the past, transformer experts have provided CRS with detailed descriptions of numerous "simple" ways terrorists could destroy HV transformers. General transformer sabotage information is also available on the Internet. One sabotage manual associated with white supremacist groups available online includes the following discussion:

> The power generation and distribution systems of most major Western cities are surprisingly vulnerable.... Attacking during peak consumption times (Winter in cold climates and Summer in hot climates) will make power diversion impossible.... Arson, explosives or long-range rifle fire can be used to disable substations, transformers and suspension pylons. A simultaneous attack against a number of these targets can shut down power ... with the advantage that service cannot be quickly restored by diverting power from another source. Each broken link in the power grid must be repaired in order to fully restore service. An individual, equipped with a silenced rifle or pistol, could easily destroy dozens of power transformers in a very short period of time.[35]

Security analysts and other industry officials acknowledge that the vulnerability of HV transformers in general is widely known, although understanding the criticality of particular assets within the power grid would require more dedicated effort.

Physical Security Measures for HV Transformers

Although HV transformers are relatively large and often exposed, frequently in rural areas, there are a number of measures available to help prevent an intentional physical attack against a transformer substation. Many of these measures are employed for public safety and to protect against theft, so they may serve multiple purposes. Although security measures appropriate for a particular substation vary depending upon its particular configuration and operating profile, such measures fall into a set of general categories:

- **Protecting information** about critical HV substations, such as engineering drawings, power flow modeling runs, and site security information, which could be useful to a potential attacker.
- **Surveillance and monitoring** through the use of video cameras, motion detectors, imaging, acoustical monitors, aerial drones, and periodic inspection by security employees.
- **Restricting physical access,** such as limiting entry only to necessary employees, installing electronic locks and other access controls, and erecting physical barriers and controls for vehicle entry. Posting full-time guards may also be an option in some circumstances.
- **Shielding assets** from offsite attacks using visual barriers such as opaque or hardened fencing, erecting taller fences, or erecting protective walls.
- **Modifying substation designs** to make them more resistant to physical damage, for example, by strengthening transformer cooling systems or bushings.
  Reconfiguring substation layouts to limit asset visibility or limit the spread of fire may also be options.

Industry and federal efforts to promote the deployment of such physical security measures are discussed later in this report. In addition to these categories, other measures can help to mitigate the immediate effects of a successful attack ("resiliency"), or to speed full system recovery from such an attack. Measures to enhance the cybersecurity of substation information and control systems, especially supervisory control and data acquisition (SCADA) systems are an important component of power grid security and are usually coordinated with physical security measures.

SECTOR INITIATIVES FOR HV TRANSFORMER SECURITY

Over the last decade or so the electric utility industry and government agencies have engaged in a number of initiatives to secure HV transformers from physical attack and to improve recovery in the event of a successful attack. These initiatives include coordination and information sharing, spare equipment programs, security standards, grid security exercises, and other measures discussed below.

Coordination and Information Sharing

The *National Infrastructure Protection Plan* (NIPP), initially published by the Department of Homeland Security in 2006, "outlines how government and private sector participants in the critical infrastructure community work together to manage risks and achieve security and resilience outcomes."[36] The plan organizes critical infrastructure into distinct sectors, designating a federal department or agency as the lead coordinator for each sector—the Sector Specific Agency (SSA). Under the NIPP and Presidential Policy Directive 21 on Critical Infrastructure Security and Resilience, the Department of Energy (DOE) is designated as the SSA for the Energy Sector, which includes the electric utility industry (excluding nuclear power plants). As an SSA, the department is responsible for working with the Department of Homeland Security (DHS), other federal agencies, critical infrastructure owners, independent regulators, and other agencies to implement national policy on critical infrastructure security and resilience.[37] The NIPP also establishes a sector partnership model including private and government coordinating councils:

- The **Electricity Subsector Coordinating Council (ESCC),** initially established in 2004, was organized and administered by companies in the electric power industry to meet regularly to coordinate policy-related activities designed to "improve the reliability and resilience of the electricity subsector, including physical and cyber infrastructure."[38] Through August 15, 2013, the ESCC was chaired by the North American Electric Reliability Corporation (NERC), the not-for-profit organization responsible for ensuring the reliability of the North American grid.[39] The ESCC has since transitioned to a new structure led by electric utility industry executives, although NERC's chief executive officer remains on the ESCC steering committee.[40]

- The **Energy Sector Government Coordinating Council (EGCC)**, also established in 2004, is the government counterpart to the ESCC. The EGCC is chaired by the DOE and DHS, incorporating other agencies at all levels of government with interest in energy security. The EGCC plays a key role in implementing the Sector-Specific Plan (discussed below), collaborating with the ESCC to develop and prioritize security programs and initiatives.[41]

In addition to these councils, other organizations have been established with more specific responsibilities related to grid security.

- The **Electricity Sector Information Sharing and Analysis Center (ES-ISAC)**, established in 1998, is the electricity sector's primary communications channel for security-related information, situational awareness, incident management, and coordination.[42] The ES-ISAC is operated by NERC in collaboration with the DOE and ESCC. Members may anonymously share security-related incident information with the ES-ISAC by means of a secure Internet portal. Registered users receive information on security threats and alerts, remediation, task forces, events, and other security-specific resources.[43]
- NERC's **Critical Infrastructure Protection Committee (CIPC)** coordinates NERC's security initiatives and advises NERC's Board of Trustees, its standing physical and cybersecurity committees, and the ES-ISAC. One of the CIPC's key functions is developing, reviewing, and revising security guidelines; and assisting in the development and implementation of NERC standards.[44]

DOE's Energy Sector-Specific Plan

The 2006 *National Infrastructure Protection Plan* required each critical infrastructure sector to develop a Sector-Specific Plan (SSP) that describes strategies to protect its critical infrastructure, outlines a coordinated approach to strengthen its security efforts, and determines appropriate funding for these activities. The section of the DOE's *Energy Sector-Specific Plan* addressing electricity was developed in collaboration with the ESCC and EGCC. The plan identifies high-voltage transformers as an electric sector vulnerability due to their criticality to the power grid and the difficulty of replacing them in the event of a successful attack. Among other measures, the SSP established a goal of implementing "agreements that require participants to maintain transformers for possible sharing in the event of a terrorist act."[45] The plan also identified the "need for a new type of emergency spare (recovery/mobile) high-voltage transformer that can be deployed and energized quickly to rapidly recover from outages caused by natural disasters and deliberate attacks."[46]

ESCC's Critical Infrastructure Strategic Roadmap

In November 2010, the Electricity Subsector Coordinating Council published its *Critical Infrastructure Strategic Roadmap* report, to provide a framework for identifying risks that could seriously disrupt the grid and for promoting actions to enhance grid reliability and resilience. The report paid particular attention to "severe-impact risks with the potential to impact large portions of the grid, or disrupt service for an extended period of time."[47] The report considered three principal risk scenarios, including

> **Scenario 1: Physical Attack on Significant Electricity System Equipment**
> A coordinated physical attack on key nodes of the bulk power system critically disables difficult to replace equipment in multiple generating stations or substations and could have a significant affect [sic] on the remainder of the system. A prolonged period of time is required to fully restore the bulk power system to normal operation.[48]

The report recommended a current capability assessment to prevent and respond to such a scenario as a "high priority." The report also recommended as "important" both a study of "options and practices to enhance physical protection of critical equipment requiring long recovery times (e.g., large high-voltage transformers)" and an initiative to "enhance the availability of critical spare equipment ... starting with high voltage transformers."[49]

Transformer Equipment Programs

Consistent with the recommendations of the studies above, several programs have been instituted within the electric power sector to address the operational issues that emerge due to the scarcity of spare HV transformers and associated equipment in the event of a physical attack or other grid emergency.

DHS Recovery Transformer Program

In 2008, the Department of Homeland Security (DHS) initiated a program to develop a prototype "Recovery Transformer" (RecX) which could enable recovery from transformer failure within days rather than months or longer.[50] The RecX transformer was intended to be adaptable to a range of common grid specifications as well as being smaller, lighter, easier to transport, and quicker to install than conventional HV transformers. The RecX prototype was designed to replace the most common HV transformers (345 kV) used in the U.S. grid.[51] This configuration reportedly could be used to replace approximately one quarter of the 2,100 transformers in this voltage class currently deployed.[52] In 2012, the only three single-phase RecX prototype units were installed in an operating 345 kV substation in Texas during a simulated emergency drill. The units remain in operation, having met or exceeded their service requirements. Although the RecX transformers have reliability and efficiency characteristics comparable to other 345 kV transformers, and are also comparably priced ($7.5 million each), the manufacturer had received no orders for commercial production of these units as of February 2014.[53] Having successfully demonstrated the RecX concept, the DHS is no longer funding the RecX program.

EEI Spare Transformer Equipment Program

In 2006, Edison Electric Institute (EEI), the main trade association for U.S. investor-owned electric utilities, initiated its Spare Transformer Equipment Program (STEP) to strengthen "the sector's ability to restore the nation's transmission system more quickly in the event of a terrorist attack."[54] The STEP program requires participating utilities to maintain (or acquire) a specific number of transformers up to 500 kV to be made available to other utilities in case of a critical substation failure. Sharing of transformers is mandatory based on a binding contract subject to a "triggering event"—a coordinated act of deliberate, documented terrorism resulting in the destruction or disabling of a transmission substation and the declaration of a state of emergency by the President.[55] FERC granted blanket authorization for the transfer and cost recovery of transmission equipment under the STEP program in September 2006.[56] State regulators with jurisdiction over participating utilities have also granted pre-approval for STEP transfers.[57] The program is designed to deal with terrorist events, but it also provides a mechanism for voluntary sharing of transformers in other emergencies, although these may require additional regulatory approvals. EEI requires annual recertification and conducts a STEP program drill every summer to ensure the program and its members will be fully prepared to respond in the event of an actual triggering event.[58]

NERC Spare Equipment Database

In 2012, NERC initiated its Spare Equipment Database (SED) program intended to serve as a tool to "facilitate timely communications between those needing long-lead time equipment damaged in a [High Impact, Low Frequency] event and those equipment owners who may be able to share existing equipment being held as spares by their organization."[59] The SED program is a confidential web-based catalog of spare transformers rated at 100 kV or higher. Only NERC and the equipment owners can see their spares data (although NERC can make high-level reports to FERC); requests for equipment are double-blind. Participation is voluntary and requires no commitment or mandatory sharing of spares.[60] Unlike EEI's STEP program, however, the SED program has not been granted pre-approval from FERC or state regulators for equipment transfers. Thus, the ability to transfer the ownership of transformers from one company to another may require additional approvals, even during an emergency.

Grid Security Exercises and Simulations

NERC and FERC have conducted grid security computer simulations and exercises specifically incorporating hypothetical attacks on HV transformer substations.

GridEx and GridEx II

In 2011, NERC conducted GridEx 2011, its first electric sector-wide grid security exercise. The exercise assessed the readiness of utilities to respond to a cyberattack, strengthened their crisis response, and provided input for internal security program improvements. Although the exercise was focused on a cyberattack, it did involve physical incursions into power grid substations as well as aspects of grid monitoring and recovery that would be relevant to an attack on HV transformers.[61] Among other findings, the exercise determined that "utilities took appropriate steps to secure the grid."[62] Nonetheless, NERC recommended that "entities should ensure their response protocols address a coordinated threat," and that it would "facilitate and support the development of updated physical security guidance."[63]

After the Metcalf attack in 2013, NERC conducted a second, more expansive grid security exercise, GridEx II. The exercise scenario, developed using open-source techniques, included a cyberattack on the grid coupled with a coordinated physical attack against a subset of transmission and generation assets—including HV transformer substations.[64] Among other conclusions, NERC's after-action report stated:

> While the electricity industry has experienced occasional acts of sabotage or vandalism, a well-coordinated physical attack also presents particular challenges for how the industry restores power.... The extreme challenges posed by the Severe Event scenario provided an opportunity for participants to discuss how the electricity industry's mutual aid arrangements and inventories of critical spare equipment may need to be enhanced.[65]

NERC did not publicly report details about the overall impacts to the grid or outages in particular regions due to the sensitive nature of such information. Utilities and other agencies participating in the exercise viewed it a useful tool for utilities to test their readiness and preparedness for attacks on the grid.[66]

FERC "Electrically Significant Locations" Study

In early 2013, prior to the Metcalf attack, then-FERC Chairman John Wellinghoff directed FERC staff to prepare an analysis identifying critical HV substations in the North American power grid.[67] Using power flow analysis software to model the impacts to the transmission system from the loss of specific grid assets,[68] FERC staff compiled a list of "Electrically Significant Locations (ESLs)" within the grid.[69] Neither details of the ESL study methodology nor its results have been released publicly by FERC or other agencies, although some findings have been reported in the press and discussed publicly by federal officials. According to the *Wall Street Journal*, the FERC analysis identified 30 critical transformers substations; in FERC's simulation, losing nine of these substations (in various combinations) as the result of a coordinated attack reportedly was found to cause a nationwide blackout for an extended time.[70]

Members of Congress were highly critical of both the *Wall Street Journal* and FERC officials for inappropriately releasing what was perceived to be highly sensitive information about power grid physical vulnerability.[71] A subsequent investigation by the Department of Energy's Inspector General concluded that FERC's handling of the ESL study findings was improper.[72] The protection of information about grid security is further discussed in a later section of this report.

HV Transformer Security Standards

Several grid security guidelines or standards have been developed or proposed to address the physical security of the grid, including HV transformers. These standards have been promulgated by NERC as voluntary best practices since at least 2002, with subsequent revisions. However, in the wake of the Metcalf incident, FERC has ordered the imposition of mandatory physical security standards in 2014.

IEEE Substation Security Standard

In 2000, the Institute of Electrical and Electronics Engineers (IEEE), a technical professional society, published its first standards for electric power substation physical and electronic security. The voluntary standard addressed "security issues related to human intrusion upon electric power supply substations" and various methods to mitigate them.[73] The standard called for the development of security assessments and, for "high-risk areas," increased security measures such as motion detectors, perimeter/area detection systems, security cameras, physical barriers, and posted guards.[74] However, according to the IEEE, the standard is intended to address security issues related to unauthorized access, theft, and vandalism. The IEEE states that "attacks against the substation for the purpose of destroying its capability to operate, such as explosives, projectiles, vehicles, etc. are beyond the scope of this standard."[75]

NERC Physical Security Guidance

In June 2002, NERC published its initial guidance for physical response to security alerts from the federal government. This alert system was revised in October 2002 to correspond to DHS's new color-coded threat level system.[76]

NERC's guidance was voluntary, intended to provide "examples of security measures that electric utility organizations may consider taking, based on the Alerts issued."[77] NERC's guidance included 35 specific security measures for the five threat DHS levels. These measures ranged from "occasional" workforce awareness programs and annual security plan reviews during times of low threat (green) to continuous monitoring of critical facilities, potentially with armed guards, during times of highest threat (red).[78] Along with this guidance, NERC published initial guidelines for vulnerability and risk assessment to help identify critical facilities and countermeasures to mitigate threats.[79]

In November 2005, NERC published a third version of its physical security guidelines, to provide "examples of security measures that other electricity sector organizations *should* consider when responding to threat level alerts" [emphasis added].[80] Thus, while still voluntary, these measures appear to have been intended as recommendations rather than considerations as stated in the earlier versions. The 2005 document included 55 measures, including new measures and existing measures expanded or described more specifically. New measures during times of low threat included, for example, annual audits of critical facility access programs and identifying critical facility long-term and short-term security measures (e.g., vulnerability assessments and security barriers).[81]

The Energy Policy Act of 2005 (P.L. 109-58) mandated the implementation of electric grid reliability standards under new authority granted to the Federal Energy Regulatory Commission. FERC subsequently designated NERC as the Electric Reliability Organization certified by the commission to establish and enforce reliability standards for the U.S. electric transmission grid, subject to commission review. In 2008, FERC approved NERC's initial reliability standards for critical infrastructure; however, these standards were developed primarily to address transmission grid cybersecurity, not physical security.[82] Subsequent NERC standards have expanded these cybersecurity requirements.

In October 2013, NERC published its most recent revision to its physical security guidance, *Security Guideline for the Electricity Sub-sector: Physical Security Response,* providing to electricity sector members "actions they should consider when responding to the threat alerts" issued by the DHS.[83] Continuing its voluntary (rather than regulatory) approach to physical security, NERC's guidance states that "each organization decides the risk it can accept and the practices it deems appropriate to manage its risk."[84] This version of NERC's guidance lays out 77 distinct security measures corresponding to three levels of threat: (1) Normal Operations/Best Practices, (2) Elevated, and (3) Imminent.

FERC Physical Security Best Practices

In 2013, FERC staff along with staff from the Federal Bureau of Investigation (FBI), DOE, DHS, and NERC participated in a number of meetings with utilities and law enforcement agencies to discuss immediate findings and recommendations stemming from the Metcalf substation attack. As part of these meetings, FERC staff shared with utilities a list of best practices for physical security. Although the list has not been made public, it reportedly included prescriptive security measures (e.g., outward-facing video surveillance) focused on security threats similar to that experienced at the Metcalf substation.[85] In 2014, DHS, in coordination with FERC, the ES-ISAC, NERC, the FBI, and industry experts, has convened another series of regional briefings across North America with utilities and law enforcement officials to follow up on the initial outreach regarding substation physical security.[86]

NERC Physical Security Regulations

On March 7, 2014, FERC ordered NERC to submit to the commission within 90 days proposed reliability standards requiring certain transmission owners "to take steps or demonstrate that they have taken steps to address physical security risks and vulnerabilities related to the reliable operation" of the power grid.[87] In its order FERC states that physical security standards are necessary because "the current Reliability Standards do not specifically require entities to take steps to reasonably protect against physical security attacks."[88] According to FERC's order, the new reliability standards must require transmission owners or operators to perform a risk assessment of their systems to identify their "critical facilities," evaluate the potential threats and vulnerabilities to those identified facilities, and develop and implement a security plan designed to protect against attacks to those identified critical facilities.[89] The order requires that each of these steps be verified by NERC or another third party qualified to review them.

On May 23, 2014, NERC filed with FERC its proposal for mandatory physical security standards.[90] The proposed standard applies to transmission owners with assets operating at 500 kV or higher as well as owners with substations operating between 200 kV and 499 kV if they meet certain interconnection or load-carrying criteria.[91] The standard consists of six principal requirements (R1-R6), summarized as follows:

R1. Risk assessments by transmission owners to identify critical transmission facilities;

R2. Independent third party verification of risk assessments conducted under R1;

R3. Requirement for transmission owners with critical facilities identified under R1 but not under their operational control to notify the transmission operator of these facilities;[92]

R4. Mandatory threat and vulnerability assessments for critical facilities conducted by transmission owners and operators;

R5. Development, documentation, and implementation of physical security plans to protect critical facilities; and

R6. Independent third party review of the threat and vulnerability assessments performed under R4 and security plans developed under R5.[93]

The proposed standard also lays out a process for compliance monitoring and assessment including audits, self-certifications, spot checking, violation investigations, self-reporting, and handling complaints.[94] The new standard would be enforced by NERC or another Regional Entity under a penalty review policy for mandatory reliability standards approved by FERC subject to the Commission's enforcement authority and oversight under P.L. 109-58.[95]

Company-Specific Initiatives

Electric utilities have long had an ongoing responsibility to ensure grid reliability, in part through operating practices and investments related to grid safety and security.[96] As the standards in the previous section suggest, there has been some level of physical security investment and an increasing refinement of grid security practices across the electric power sector for at least the last 15 years. Nonetheless, several major transmission owners have recently announced significant new initiatives specifically to improve the physical security of critical transformer substations in light of the Metcalf attack. Other utilities have included new substation security investments in broader initiatives for company security.[97] The following examples illustrate the types of security changes being proposed by these grid owners. Note that other major utilities have not publicly announced similar new security initiatives. A comprehensive review or comparison of physical security plans among all major grid owners in the United States is beyond the scope of this report.

The Tennessee Valley Authority

In February 2012, the Tennessee Valley Authority (TVA) announced that it was "realigning its operations and structure to enhance security at TVA's non-nuclear power facilities ... focusing more of our non-nuclear security resources on our critical infrastructure," including HV substations.[98] The realignment included ending uniformed patrols in favor of installing more security technology, and the stationing of contract guards 24 hours a day at critical facilities. Together with local law enforcement cooperation, the shift to contract guards was intended to provide a more persistent security presence and faster incident response at key locations. Among the security technologies reportedly deployed by TVA are "surveillance, infrared cameras, video analytics for alarm verification and assessment, virtual perimeters, card readers, [and] automated gates."[99] TVA's security initiatives in 2012 appear to have been motivated primarily by security concerns such as copper theft, but would be applicable to more serious security risks such as terror attacks. In February 2014, after the Metcalf incident, TVA reportedly stated that it was "intensifying efforts" to educate local law enforcement about the importance of substations, including taking police on site visits to see substations during normal operations.[100] The utility has also been canvassing residents near TVA property asking them to report unusual activity around grid facilities.

Pacific Gas and Electric (PG&E)

In February 2014, in response to the attack on its Metcalf substation, PG&E announced that it would be investing approximately $100 million over three years to improve substation security. Physical security measures mentioned by the company include new perimeter barriers, shielding for certain equipment, more cameras (inside and outside the fence), and clearing vegetation. For its most critical facilities, the company is "studying advanced detection technology such as night vision and thermal imaging."[101] Other security measures mentioned in news reports about PG&E include enhanced lighting, 24-hour security guards, and increased patrols by local law enforcement agencies.[102]

Dominion

In February 2014, Dominion Virginia Power, an operating company of Dominion, announced plans to spend up to $500 million over five to seven years "to harden its transmission substations and other critical infrastructure against man-made physical threats and natural disasters, as well as stockpile crucial equipment for major damage recovery."[103] Dominion reportedly began to increase substation security efforts in 2013, focusing first on substations at greatest risk.[104] Among the security measures identified by the utility are physical barriers, additional access control, equipment design/hardening, polymer bushing installation, additional spare equipment, and relocation of spare equipment to off-site storage areas. Other measures reportedly include dual-perimeter "no man zones" around substations and installing systems for key-card access to substation yards.[105] Dominion's security plan has yet to be approved by Virginia regulators for cost recovery in electric rates.

Bonneville Power Administration

In its 2014 draft *Security Asset Management Strategy*, the Bonneville Power Administration (BPA) proposes approximately $37 million in additional capital spending through FY2020 for physical security measures at approximately 60 critical transformer substations.[106] BPA's *Strategy* states that, over the last 13 years, the utility "has conducted hundreds of security and risk assessments using several industry accepted methodologies," and began implementing security improvements based on these risk assessments beginning in 2001.[107]

ISSUES FOR CONGRESS

The recent transformer substations attacks, together with federal grid security exercises, have focused attention on the vulnerability of HV transformer substations to organized physical attacks. As the electric power industry and federal agencies continue their efforts to improve the physical security of critical HV transformer substations, Congress may consider several key issues as part of its oversight of the sector.

Identifying Critical Transformers

A fundamental consideration regarding HV transformer security is a clear and stable understanding of which transformers are "critical." The USA PATRIOT Act of 2001 defines "critical infrastructure" in the most general sense as "systems and assets ... so vital to the United States that the incapacity or destruction of such systems and assets would have a debilitating impact on security, national economic security, national public health or safety, or any combination of those matters."[108] It its 2009 guidelines for identifying critical assets specifically in the electricity sector, NERC defines critical assets as those "that if destroyed, degraded, compromised (e.g., misused) or otherwise rendered unavailable would unacceptably affect the reliability or operability of the [Bulk-Power System] as a whole.... "[109] FERC's 2014 order mandating physical security standards for the grid defines a "critical facility" as "one that, if rendered inoperable or damaged, could have a critical impact on the operation of the interconnection through instability, uncontrolled separation or cascading failures on the Bulk-Power System."[110] All three definitions associate "criticality" with a failure event of national significance, although none provides a more prescriptive basis for identifying such assets.

In its physical security order, FERC does not require that a "mandatory" number of critical facilities be identified under the standards.[111] Determination of whether a specific HV transformer is "critical" will be based on each individual asset owner's "objective analysis, technical expertise, and experienced judgment."[112] In its proposed physical security standards, NERC requires transmission owners with HV assets meeting prescriptive criteria to examine whether they *may* have critical transformers, but it is up to the owners to determine themselves if any of their assets *are* critical through a periodic risk assessment based on their own respective transmission analyses, subject to independent validation.[113] Thus, grid owners could have considerable latitude in determining which of their transformer substations (if any) are critical and therefore subject to the requirements of the new standard.

Although there are many candidate transformer substations in the grid, relatively few are likely to be of national significance. As discussed above, of the numerous HV transformer substations in the United States, FERC's 2013 power flow analysis identified only 30 as being critical to the national grid (although each of these substations may contain multiple HV transformers). Whether the number of critical transformer substations under FERC's definition above turns out to be higher or lower than 30, it will likely be only a small fraction of the total asset base. This conclusion is consistent with FERC's expectation that under NERC's new standard "the number of facilities identified as critical will be relatively small.... For example, of the many substations on the Bulk-Power System, our preliminary view is that most of these would not be 'critical' as the term is used in this order."[114] Consistent with this view, the NERC working group responsible for drafting the proposed physical security standard likewise expects the number of critical facilities to be "small and that many Transmission Owners that meet the applicability of this standard will not actually identify any such Facilities."[115]

Properly identifying which HV transformer substations are critical is a key issue. Otherwise, the electricity sector risks the possibility of hardening too many substations, hardening the wrong substations, or both. Either outcome could increase ultimate costs to electricity consumers without commensurate security benefits, and could potentially divert limited security resources from other important grid priorities (e.g., cybersecurity). Independent verification is intended to validate utility assessments of substation criticality, but the standard's reliance on company-bycompany assessments may still allow for important differences in analytic methodology or assumptions, and thus inconsistent conclusions about transformer criticality. Furthermore, company-specific studies may not align with a "top down" assessment of asset criticality like that performed by FERC in its Electrically Significant Location (ESL) analysis. Congress may examine whether company-specific assessments of transformer criticality could differ from national-level assessments and what implications, if any, such differences might have on overall grid security and company efforts to protect particular substations.

Confidentiality of Critical Transformer Information

Ensuring the confidentially of critical infrastructure information has been a long-standing concern across all critical infrastructure sectors. It is a key reason for the establishment of sector Information Sharing and Analysis Centers (ISACs), including the Electricity Sector ISAC, discussed above. Confidentiality also factors into the administration of the industry's spare transformer programs and other activities related to critical infrastructure. FERC has established policies for the protection of critical energy infrastructure information (CEII) through a series of orders, beginning with Order 630, issued February 21, 2003.[116] The order (§ 27) defines CEII as information that "must relate to critical infrastructure, be potentially useful to terrorists, and be exempt from disclosure under the Freedom of Information Act [FOIA]." It also establishes procedures and responsibilities for determining what information qualifies as CEII and handling CEII requests.[117] FERC's 2014 order mandating physical security standards also requires procedures to ensure confidential treatment of sensitive information.[118]

Press articles in the wake of the Metcalf attacks, notably in the *Wall Street Journal*, cited specific details about FERC's 2013 ESL analysis, reportedly from a copy of a FERC presentation obtained by the paper. Notwithstanding FERC's orders on CEII, Members of Congress and FERC officials have expressed concern that the release of the presentation by FERC staff and the publication of details in the press potentially compromised grid security.[119] Others reportedly have disputed this concern, including the former FERC Commissioner responsible for commissioning and presenting the ESL study findings at industry meetings.[120] In April 2014, the DOE Inspector General concluded that the FERC presentation in question "should have been classified and protected from release" and "that the Commission may not possess adequate controls for identifying and handling classified national security information."[121] The Acting Chairman of FERC has testified that the commission is adopting the Inspector General's recommendations to improve its handling of CEII and requested additional authority from Congress for exemption from FOIA.[122]

FERC staff may be improving the way CEII is safeguarded in response to the Inspector General's report, but securing CEII may continue to be an issue if NERC's new physical security regulations are approved by the commission. NERC's regulations would require independent risk assessments by multiple grid owners and 3rd party validation of those assessments. This process, by construction, would cause considerable new CEII to be created (e.g., multiple Midwest power flow models) and shared among utilities, RTOs, and consultants in ways that may be new to the industry. Ensuring that CEII generated and transferred among these entities remains secure could require special attention. As FERC's improper management of the ESL study information shows, having strong CEII policies in place may not guarantee that those policies will be correctly and uniformly followed—even by the agency that created them.

Adequacy of HV Transformer Protection

The electric power sector has had physical security guidelines in place for well over a decade, as discussed above. These voluntary guidelines have been updated and expanded periodically to reflect industry experience, changes in the security environment, and new technologies. Prior to 2014, however, it appears that the physical security initiatives among grid owners were focused primarily on preventing vandalism and theft (of copper wire) rather than a terrorist attack.[123] As the recent substation attacks in California, Arkansas, and Florida have shown, many other security measures available to grid owners were not implemented—even at critical HV substations.

A grid owner's focus on vandalism and theft may be understandable because such incidents have occurred frequently and their associated costs are tangible and well-understood. Investing in security against a terrorist attack presents a greater challenge in terms of costs and benefits. As a 2006 report from the Electric Power Research Institute states,

> Security measures, in themselves, are cost items, with no direct monetary return. The benefits are in the avoided costs of potential attacks whose probability is generally not known. This makes cost-justification very difficult.[124]

Note that cost-justification requires not only the approval of utility management, but also of FERC and potentially state public utility commissions which regulate the rates grid owners may charge for electric transmission and distribution service. Regulators are responsible for ensuring that electricity rates are just and reasonable. They must be convinced that any new grid security capital costs and expenses are necessary and prudent before they will allow them to be passed through to ratepayers.

The Metcalf incident and GridEx exercises have provided the electric sector with valuable new information about the potential threat, vulnerability, and consequence of a coordinated attack on HV transformers. Risk assessments incorporating this information presumably would justify (with or without a new NERC standard) increased security investments at critical substations to prevent intentional attacks. The recently announced voluntary spending plans at PG&E, Dominion, and BPA for HV substation security appear to reflect such risk and cost-benefit reassessments. Nonetheless, there continues to be considerable uncertainty about the risk of terror attacks on the power grid, and what measures are economically justified in addressing them. PG&E, BPA, and the other utilities announcing large security investments have already decided to make such investments, but they are in the minority. Other major owners of critical HV transformers have not publicly announced similar plans.

NERC's proposed standards for power grid physical security would ensure considerable consistency in the *process* utilities must undertake to identify critical substations and develop plans to secure them. However, they may not ensure consistency among the various security plans nor in the specific measures the individual asset owners will choose to implement to reduce the risk of intentional attacks. As FERC continues to implement its policy of regulating physical security of the power grid, Congress may examine whether company-specific security initiatives appropriately reflect the risk profiles of their particular assets, and whether additional security measures across the grid overall uniformly reflect terrorism risk from a national perspective.

Quality of Federal Threat Information

The power industry's physical security risk assessments rely upon information about security threats provided by the federal government, among other sources, communicated through the ISAC, during DHS and other agency briefings, or through other channels. The quality of this threat information is a key determinant of what grid owners need to be protecting against and what security measures to take. Incomplete or ambiguous threat information— especially from the federal government—may lead to inconsistency in physical security among grid owners, inefficient spending of limited security resources at facilities (e.g., that may not really be under threat), or deployment of security measures against the wrong threat. For example, prior to FERC's physical security order, the head of NERC, which initially opposed mandatory physical security standards stated,

> I am concerned that a rule-based approach for physical security would not provide the flexibility needed to deal with the widely varying risk profiles and circumstances across the North American grid and would instead create unnecessary and inefficient regulatory burdens and compliance obligations.[125]

Differences in the interpretation or application of threat information, as discussed in the previous section, may be a reason why some large utilities have announced major new substation security initiatives while others have not.

Concerns about the quality and specificity of federal threat information have long been an issue across all critical infrastructure sectors.[126] Threat information continues to be an uncertainty in the case of power grid physical security. For example, some federal officials reportedly have characterized the Metcalf incident as a domestic terrorist attack, potentially a "dry run" for a more destructive attack on multiple HV transformer substations, while the FBI has stated that it does not believe Metcalf was a terrorist incident.[127] Because the perpetrators have not been identified, it is impossible to know for certain, but the ambiguity has significant implications for HV substation security going forward. Although there is wide consensus that the Metcalf attack was extremely serious, some industry analysts have opined that FERC's physical security order may be an "overreaction" to Metcalf.[128] By contrast, former DHS Secretary Michael Chertoff has predicted that "the sophistication and resulting damage of the Metcalf attack will ... be exceeded" in a future attack.[129] Still others have expressed concern that FERC's physical security concerns may be too heavily focused on another Metcalf-type scenario (the last threat) rather than a wider range of potential future threats (the next threat).[130]

There is widespread agreement among government, utilities, and manufacturers that HV transformers in the United States are vulnerable to terrorist attack, and that such an attack potentially could have catastrophic consequences. But the most serious, multi-transformer attacks would require acquiring operational information and a certain level of sophistication on the part of potential attackers. Consequently, despite the technical arguments, without more specific information about potential targets and attacker capabilities, the true vulnerability of the grid to a multi-HV transformer attack remains an open question. As Congress seeks to establish the best policies to address HV transformer vulnerability relative to other infrastructure security priorities, understanding this vulnerability in the context of specific demonstrable threats may become increasingly important. To this end Congress may examine how federal threat information is developed and used by grid owners, and how limitations and uncertainty of this information may affect the HV transformer physical security among electric utilities.

Recovery from HV Transformer Attacks

Physical security for HV transformer substations has the primary purpose of preventing successful attacks against these critical assets within the power grid. However, in the event of a successful attack, measures to minimize its effect on the overall grid are equally important so that the loss of any particular transformer remains a local event. To this end the electric power industry emphasizes its strategy of "defense-in-depth," which includes incident response and recovery in addition to preparation and prevention.[131] Industry initiatives to enhance grid resiliency, including incident recovery programs such as the DHS recovery transformer program and EEI's spare transformer program, contribute to the power grid's ability to sustain a terrorist attack without widespread grid failure. Indeed, some analysts have pointed to the Metcalf incident as a successful demonstration of grid resiliency; electric service was not interrupted despite the loss of a critical substation in the San Francisco Bay area. As Congress continues its examination of physical security policy, maintaining a holistic perspective on prevention and recovery as integrated aspects of HV transformer security may help to clarify an effective balance in terms of industry investment and regulatory oversight.

End Notes

[1] Portions of this report were drawn from CRS Report R42795, *Electric Utility Infrastructure Vulnerabilities: Transformers, Towers, and Terrorism*, by Amy Abel, Paul W. Parfomak, and Dana A. Shea.

[2] 1 kV=1,000 volts.

[3] North American Electric Reliability Corporation, "Understanding the Grid," fact sheet, August 2013, http://www.nerc.com/AboutNERC/Documents/Understanding%20the%20Grid%20AUG13.pdf. Note that there is no industry consensus as to what voltage rating or other operating characteristic constitutes "high voltage." This report uses 230 kV as the high voltage threshold, but other studies may use a different threshold, such as 115/138 kV, or may include an additional "extra high voltage" category above 345 kV. See, for example, U.S. Department of Energy, *Large Power Transformers and the U.S. Electric Grid*, April 2014, p. 4.

[4] C. Newton, "The Future of Large Power Transformers," *Transmission & Distribution World*, September 1, 1997; William Loomis, "Super-Grid Transformer Defense: Risk of Destruction and Defense Strategies," Presentation to NERC Critical Infrastructure Working Group, Lake Buena Vista, FL, December 10-11, 2001.

[5] See, for example: Senators Dianne Feinstein, Al Franken, Ron Wyden, and Harry Reid, letter to the Honorable Cheryl LaFleur, Acting Chairman, Federal Energy Regulatory Commission, February 7, 2014, http://www.ferc.gov/industries/ electric/indus-act/reliability/chairman-letter-incoming.pdf.

[6] The three currents are sine wave functions of time with the same frequency (60 Hertz). The phases are spaced equally, offset 120 degrees from each other. With three-phase power, one of the phases is always nearing a peak.

[7] The loss of power on the transmission system is proportional to the square of the current (flow of electricity) while the current is inversely proportional to the voltage.

[8] Pauwels Canada, Inc., personal communication, October 20, 2003.

[9] U.S. Department of Energy, April 2014, p. 7.

[10] U.S. Department of Energy, April 2014, p. 9.

[11] Pauwels Canada, Inc., October 20, 2003.

[12] U.S. Department of Energy, April 2014, p. 7.

[13] C. Newton, "The Future of Large Power Transformers," *Transmission & Distribution World*, September 1, 1997.

[14] C. Newton, September 1, 1997.

[15] U.S. Department of Energy, April 2014, p. 27.

[16] Kenneth Friedman, U.S. Department of Energy, "DOE Update on GMD/EMP-Related Activities," Presentation to the Geomagnetic Disturbance Task Force Working Group, North American Electric Reliability Corporation, November 13, 2013.

[17] John Kappenman, *Geomagnetic Storms and Their Impacts on the U.S. Power Grid*, Meta-R-319, Metatech Corp., prepared for Oak Ridge National Laboratory, January 2010, p. 1-14, http://www.ferc.gov/industries/electric/indus-act/reliability/cybersecurity/ferc_meta-r-319.pdf.

[18] HV substation information for specific investor-owned utilities is publicly available in annual reports filed with the Federal Energy Regulatory Commission (FERC Form-1).

[19] National Research Council (NRC), *Terrorism and the Electric Power Delivery System*, 2012, p. 69.

[20] Office of Technology Assessment (OTA), *Physical Vulnerability of Electric Systems to Natural Disasters and Sabotage*, OTA-E-453, June 1990, p. 37.

[21] See, for example, Réka Albert, István Albert, and Gary L. Nakarado, "Structural Vulnerability of the North American Power Grid," *Physical Review E*, Vol. 69, 025103(R), 2004.

[22] Rebecca Smith, "U.S. Risks National Blackout From Small-Scale Attack," *Wall Street Journal*, March 12, 2014.

[23] Rebecca Smith, "Assault on California Power Station Raises Alarm on Potential for Terrorism," *Wall Street Journal*, February 5, 2014.

[24] Mitsubishi Electric Power Products, Inc., personal communication, Warrendale, PA, September 23, 2003.

[25] Lower Colorado River Authority, "August 6 Update on Transformer Fire," ,ress release, Austin, TX, August 6, 2003.

[26] RTO Insider, "Substation Saboteurs 'No Amateurs'," April 2, 2014, http://www.rtoinsider.com/pjm-grid2020-1113- 03/.

[27] Chelsea J. Carter, "Arkansas Man Charged in Connection with Power Grid Sabotage," CNN, October 12, 2013; Max Brantley, "FBI Reports Three Attacks on Power Grid in Lonoke County," *Arkansas Times*, October 7, 2013.

[28] Jim Peppard, "Reward Offered in Power Transformer Shooting," WTSP News (Tampa), October 17, 2005.

[29] David Egan and Kenneth Seiler, PJM Interconnection, "PJM Manages Aging Transformer Fleet," *T&D World*, March 1, 2007.

[30] Tom Daspit, "Schnabel Cars in Service," web page, August 15, 2013, http://southern.railfan.net/schnabel/ schnabel_cars.html.

[31] Marija Ilic, Professor, Engineering and Public Policy and Electrical and Computer Engineering, Carnegie Mellon Univ., Pittsburgh, PA, personal communication, September 22, 2003.

[32] NRC, 2012, p.79.

[33] OTA, June 1990, pg. 37.

[34] Stewart Tendler, "IRA Bombers Plotted to Black Out London and South East for Months," *The Times*, London, England, April 12, 1997.

[35] Axl Hess (a.k.a. Aquilifer), *White Resistance Manual V2.4*, 2001. See also Herschel Smith, "A Terrorist Attack That America Cannot Absorb," captainsjournal.com, blog, September 28, 2010, http://www.captainsjournal.com/2010/09/ 28/a-terrorist-attack-that-america-cannot-absorb/.

[36] Department of Homeland Security (DHS), "National Infrastructure Protection Plan," web page, April 7, 2014, https://www.dhs.gov/national-infrastructure-protection-plan. The NIPP was mandated under Homeland Security Presidential Directive 7 issued on December 17, 2003.

[37] Presidential Policy Directive 21, *Presidential Policy Directive—Critical Infrastructure Security and Resilience*, February 12, 2013.

[38] North American Electric Reliability Corporation (NERC), "Electricity Sub-sector Coordinating Council," web page, April 7, 2014, http://www.nerc.com/pa/CI/Pages/ESCC.aspx.

[39] Among other functions, NERC develops and enforces reliability standards, monitors the grid, and trains industry personnel. In the United States, NERC is subject to FERC oversight.

[40] Gerry W. Cauley, North American Electric Reliability Corporation (NERC), letter to U.S. Secretary of Energy Ernest Moniz, August 23, 2013, http://www.publicpower.org/files /PDFs/DOESecLetterHistoryESCC.pdf.

[41] Department of Energy, *Energy Sector-Specific Plan*, 2010, p. 20.

[42] The ES-ISAC was established under Presidential Decision Directive 63, May 22, 1998.

[43] Electricity Sector Information Sharing and Analysis Center (ES-ISAC), "Frequently Asked Questions," web page, https://www.esisac.com/SitePages/FAQ.aspx.

[44] North American Electric Reliability Corporation (NERC), "Critical Infrastructure Protection Committee (CIPC)," web page, http://www.nerc.com/comm/CIPC/Pages/default.aspx, April 8, 2014.

[45] Department of Homeland Security, *Energy Sector-Specific Plan*, 2010, p. 54.

[46] Department of Homeland Security, *Energy Sector-Specific Plan*, 2010, p. 70.

[47] North American Electric Reliability Corporation (NERC), *Critical Infrastructure Strategic Roadmap*, November 2010, p. 2, http://ccpic.mai.gov.ro/docs/NERC_ESCC_Critical_ Infrastructure_Strategic_Roadmap.pdf.

[48] NERC, November 2010, p.18. This scenario involved the loss of three HV substations serving large urban centers with a restoration time to 100% operating capacity of 6-18 months.

[49] NERC, November 2010, pp.19-20.

[50] The program was partly funded by the DHS Science and Technology Directorate in a consortium with the Electric Power Research Institute, CenterPoint Energy, and ABB.

[51] ABB, "US Rapid Recovery Transformer Initiative Succeeds Using Specially-Designed ABB Transformers," press release, October 4, 2012.

[52] Matthew L. Wald, "A Drill to Replace Crucial Transformers (Not the Hollywood Kind)," *New York Times*, March 14, 2012.

[53] National Research Council (NRC), *The Resilience of the Electric Power Delivery System in Response to Terrorism and Natural Disasters: Summary of a Workshop*, 2013; Sarah Mahmood, Department of Homeland Security, personal communication, February 10, 2014.

[54] Edison Electric Institute (EEI), "Spare Transformers," web page, April 10, 2014, http://www.eei.org/issuesandpolicy/ transmission/Pages/sparetransformers.aspx.

[55] Edison Electric Institute (EEI), "Overview of the Spare Transformer Equipment Program," slide presentation, February 23, 2014.

[56] Federal Energy Regulatory Commission, *Order on Application for Blanket Authorization for Transfers of Jurisdictional Facilities and Petition for Declaratory Order*, Docket Nos. EC06-14-000 and EL06-86-000, September 22, 2006.

[57] EEI, February 23, 2014.

[58] Edison Electric Institute, briefing for the Congressional Research Service, February 23, 2014.

[59] North American Electric Reliability Corporation (NERC), *Special Report: Spare Equipment Database System*, August 2011.

[60] North American Electric Reliability Corporation (NERC), "Spare Equipment Database," slide presentation, NERC Industry Webinar, July 22, 2013, http://www.nerc.com/ pa/RAPA/webinardl/SED_Presentation_July_22_2013.pdf.

[61] North American Electric Reliability Corporation (NERC), *2011 NERC Grid Security Exercise: After Action Report*, March 2012, p. i.

[62] NERC, 2012, p. ii.

[63] Ibid.

[64] North American Electric Reliability Corporation (NERC), *Grid Security Exercise (GridEx II): After-Action Report*, March 2014, p.15; Matthew L. Wald, "Attack Ravages Power Grid. (Just a Test.)," *New York Times*, November 14, 2013.

[65] NERC, March 2014, p. 5.

[66] See, for example, American Public Power Association, "Physical Security and the Electric Sector," fact sheet, February 2014, http://www.publicpower.org/files/PDFs /Physical SecurityIBFebruary2014.pdf; Matthew L. Wald, "Power Grid Preparedness Falls Short, Report Says," *New York Times*, March 12, 2014.

[67] Federal Energy Regulatory Commission (FERC), "Second Set of Responses of the Federal Energy Regulatory Commission to Senator Murkowski's Separately Submitted Questions for the Record from April 10, 2014 Hearing of the Senate Energy and Natural Resources Committee," May 5, 2014, pp. 12-13, http://www.energy.senate.gov/public/ index.cfm/files/serve?File_id=5c3bf9d7-bb7f-4379-8f57-f58881a0b5d6.

[68] FERC staff employed the commission's Topological and Impedance Element Ranking (TIER) model to identify "significant" assets based upon undisclosed criteria. For more details of the TIER model, see Bernard C. Lesieutre et al., "Topological and Impedance Element Ranking (TIER) of the BulkǦPower System," University of Wisconsin— Madison, prepared for the Federal Energy Regulatory Commission, August 2009, https://www.ferc.gov/EventCalendar/ Files/20090911112656-TIER%20REPORT.pdf.

[69] Federal Energy Regulatory Commission (FERC), "Response to Senator Murkowski's Separately Submitted Questions for the Record from April 10, 2014 Hearing of the Senate Energy and Natural Resources Committee, Question 39," May 5, 2014, p. 2, http://www.energy.senate.gov/public/index.cfm/files/serve?File_id=2826f80a-a986- 45d1-9261-87b45e1d6872.

[70] Rebecca Smith, "U.S. Risks National Blackout from Small-Scale Attack on Substations," *Wall Street Journal*, March 13, 2014.

[71] Senate Committee on Energy and Natural Resources, "Landrieu, Murkowski Ask Inspector General to Examine Leaks of Grid Vulnerabilities," press release, March 31, 2014.

[72] U.S. Department of Energy, Office of Inspector General, "Review of Internal Controls for Protecting Non-Public Information at the Federal Energy Regulatory Commission," DOE/IG-0906, April 9, 2014.

[73] Institute of Electrical and Electronics Engineers (IEEE), *1402-2000 - IEEE Guide for Electric Power Substation Physical and Electronic Security*, January 30, 2000.

[74] IEEE, January 30, 2000, p. 16.

[75] Institute of Electrical and Electronics Engineers (IEEE), "P1402—Standard for Physical Security of Electric Power Substations," web page, June 3, 2014, http://standards.ieee.org/develop/project/1402.html.

[76] North American Electric Reliability Corporation (NERC), *Threat Alert System and Physical Response Guidelines for the Electricity Sub-sector*, Version 2.0, October 8, 2002, http://www.iwar.org.uk/infocon/threat-levels/ tas_physical_V2.pdf.

[77] NERC, October 8, 2002, p. 2.

[78] NERC, October 8, 2002, pp. 3-4.

[79] NERC, *Security Guidelines for the Electricity Sector: Vulnerability and Risk Assessment*, June 14, 2002, http://www.esisac.com/Public%20Library/Documents/Security%20Guidelines/ Vulnerability%20and%20Risk%20Assessment,%20Version%201.0.pdf.

[80] NERC, *Security Guidelines for the Electricity Sector: Physical Response*, November 1, 2005, p.1, http://www.esisac.com/Public%20Library/Documents/Security%20Guidelines/ Physical%20Response,%20Version%203.0.pdf.

[81] NERC, November 1, 2005, p. 3.

[82] FERC Order 706.

[83] North American Electric Reliability Corporation (NERC), *NERC: Security Guideline for the Electricity Sub-sector: Physical Security Response*, October 28, 2013, p. 1, http://www.nerc.com/comm/CIPC/SecurityGuidelinesCurrent/ Electricity%20Sector%20Physical%20Security%20Guideline%20(Approved%20by%20CI PC%20- %20October%2028,%202013).pdf.

[84] NERC, October 28, 2013, p.1.

[85] Edison Electric Institute, briefing for the Congressional Research Service, February 23, 2014.

[86] Gerry Cauley, CEO, North American Electric Reliability Corporation (NERC), letter to Senator Harry Reid, February 12, 2014, p. 2, http://www.nerc.com/news/Headlines%20DL/ NERC%20Response%20to%20Senators%20Letter%20- Reid%20%202%2011%2014%20v4.pdf.

[87] Federal Energy Regulatory Commission (FERC), *Reliability Standards for Physical Security Measures*, Order Directing Filing of Standards, Docket No. RD14-6-000, March 7, 2014, p.1, http://www.ferc.gov/CalendarFiles/ 20140307185442-RD14-6-000.pdf.

[88] FERC, March 7, 2014, p. 2.

[89] FERC, March 7, 2014, pp. 3-4.

[90] North American Electric Reliability Corporation (NERC), Petition of the North American Electric Reliability Corporation for Approval of Proposed Reliability Standard CIP-014-1, May 23, 2014, http://www.nerc.com/ FilingsOrders/us/NERC%20Filings%20to%20FERC %20DL/Petition%20-%20Physical%20Security%20CIP-014- 1.pdf.

[91] NERC, May 23, 2014, Exhibit A, p. 1.

[92] A regional transmission operator (RTO) administers the transmission grid for multiple transmission owners in a specified region in accordance with FERC Order No. 2000. RTOs and independent system operators (ISOs) are defined in section 3 of the Federal Power Act (16 U.S.C. 796).

[93] NERC, May 1, 2014, Section B.

[94] NERC, May 1, 2014, p. 14.

[95] Federal Energy Regulatory Commission (FERC), *Statement of Administrative Policy on Processing Reliability Notices of Penalty and Order Revising Statement in Order No. 672*, Docket Nos. AD08-6-000 and RM05-30-002, April 17, 2008.

[96] For example, see security discussion in Con Edison, Initial Brief on Behalf of Consolidated Edison Company of New York, Inc. in Support of a Permanent Electric Rate Increase, Before the New York State Public Service Commission, November 30, 2007, http://media.corporate-ir.net/media_files/irol/61/61493/total120507.pdf.

[97] See Southern California Edison, *Safety, Security, & Compliance (SS&C):Volume 4— Corporate Security and Business Resiliency*, 2015 General Rate Case, Before the Public Utilities Commission of the State of California, November 2013, http://www3.sce.com/sscc/law/dis/dbattach5e.nsf/0/0B9F998127246B4288257C21008148 B0/$FILE/ SCE-07%20Vol.%2004.pdf.

[98] Tennessee Valley Authority, "TVA Realigns Security to Enhance Protection at Non-Nuclear Assets," press release, February 17, 2012, http://www.tva.gov/news/releases/janmar12/tvap.html.

[99] "Addressing Cyber and Physical Risks in Modern Utility Security," *Security*, March 1, 2014, http://www.securitymagazine.com/articles/85275-addressing-cyber-and-physical-risks-in-modern-utility-security.

[100] Rebecca Smith, "U.S. Utilities Tighten Security After 2013 Attack," *Wall Street Journal*, February 9, 2014.

[101] Geisha Williams, Executive Vice President of Electric Operations, Pacific Gas and Electric Company, "PG&E Metcalf Attack: Gunfire on Substation Has Led to Greater Security," *San Jose Mercury News*, April 15, 2014.

[102] "PG&E to Spend $87M on Security to Protect Large Substations from Attack," KTVU, Oakland, CA, February 12, 2014.

[103] Dominion. "Substation Security," fact sheet, Spring 2014, https://www.dom.com/about/electric-transmission/pdf/ substation-security-soc-factsheet.pdf.

[104] Tracy Sears, "Troopers Increase Security at Virginia Substations Critical to Grid," WTVR, March 11, 2014.

[105] Peter Bacqué, "Va. Power to Spend Up to $500M on Security Plan," *Richmond Times-Dispatch*, February 8, 2014.

[106] Bonneville Power Administration (BPA), *Security Asset Management Strategy*, February 2014, p. 29, http://www.bpa.gov/Finance/FinancialPublicProcesses/CapitalInvestment Review/2014CIRDocuments/ Security%20Full%20Asset%20Strategy%20Final%20Draft.pdf.

[107] BPA, February 2014, p. 31.

[108] P.L. 107-56 § 1016(e).

[109] North American Electric Reliability Corporation (NERC), "Security Guideline for the Electricity Sector: Identifying Critical Assets," September 17, 2009, p. 1, http://www.nerc.com/fileUploads/File/Standards/Reference%20Documents/ Critcal_Asset_Identification_2009Nov19.pdf.

[110] FERC, March 7, 2014, p. 3.

[111] FERC, March 7, 2014, p. 3.

[112] FERC, March 7, 2014, p. 3.

[113] NERC, May 1, 2014, p. 30.

[114] FERC, March 7, 2014, p. 3.

[115] NERC, May 1, 2014, p. 28.

[116] For an overview, see Federal Energy Regulatory Commission (FERC), "Critical Energy Infrastructure Information (CEII) Regulations," web page, June 28, 2010, http://www.ferc.gov/legal//maj-ord-reg/land-docs/ceii-rule.asp.

[117] Federal Energy Regulatory Commission (FERC), Order No. 630, Final Rule, February 21, 2003, http://elibrary.ferc.gov/idmws/common/opennat.asp?fileID=9639612.

[118] FERC, March 7, 2014, p. 10.

[119] Senate Committee on Energy and Natural Resources, "Sens. Landrieu, Murkowski Ask Inspector General to Examine Leaks of Grid Vulnerabilities," press release, March 27, 2014; The Honorable Cheryl LaFleur, Chairman (Acting), Federal Energy Regulatory Commission (FERC), Testimony Before the Senate Committee on Energy and Natural Resources Hearing, "Keeping the Lights On—Are We Doing Enough to Ensure the Reliability and Security of the U.S. Electric Grid?," April 10, 2014.

[120] Bobby McMahon, "Wellinghoff Says FERC Analysis of Grid Vulnerability was Public, Calls Review 'Waste of Time'," *Inside FERC*, March 31, 2014, p. 1.

[121] U.S. Department of Energy, Office of Inspector General, "Review of Internal Controls for Protecting Non-Public Information at the Federal Energy Regulatory Commission," DOE/IG-0906, April 2014, p. 1.

[122] The Honorable Cheryl LaFleur, Testimony on April 10, 2014.

[123] See, for example, Michael Wills, "Changes at Duke Energy Substations Crack Down on Copper Thieves," *WUNC Radio 91.5*, May 22, 2013; Scott Kraus, "Hit Hard by Copper Wire Thieves, PPL Fights Back," *The Morning Call* (Lehigh, PA), June 6, 2013.

[124] Electric Power Research Institute (EPRI), *Technologies for Remote Monitoring of Substation Assets: Physical Security*, March 2006, p. viii.

[125] Gerry Cauley, President and CEO, North American Electric Reliability Corporation (NERC), Letter to Senate Majority Leader Harry Reid, February 12, 2014, p. 2, http://www.nerc.com/news/Headlines%20DL/ NERC%20Response%20to%20Senators%20Letter%20- Reid%20%202%2011%2014%20v4.pdf.

[126] See, for example, Philip Shenon, "Threats and Responses: Domestic Security," *New York Times*, June 5, 2003, p. A15.

[127] Rebecca Smith, February 5, 2014.

[128] Deborah Carpentier, "NERC Gains in Vegetation Management, Cyber and Physical Security, and Reliability Assurance," *Natural Gas & Electricity* (Wiley Periodicals), May 2014, p. 31, http://www.crowell.com/files/NERCGains-in-Vegetation-Management-Cyber-and- Physical-Security-and-Reliability-Assurance.pdf.

[129] Michael Chertoff, "Building a Resilient Power Grid," *Electric Perspectives*, May/June 2014, p. 35.

[130] Edison Electric Institute, briefing for the Congressional Research Service, February 23, 2014.

[131] Edison Electric Institute, "The Electric Power Industry's Commitment to Protecting Its Critical Infrastructure," February 2014, http://www.eei.org/issuesandpolicy/cybersecurity /Documents/Critical_Infra_Physical_Protection.pdf.

INDEX

E

D